P9-CMP-532

# shapes, space, and symmetry

# shapes, space, and symmetry

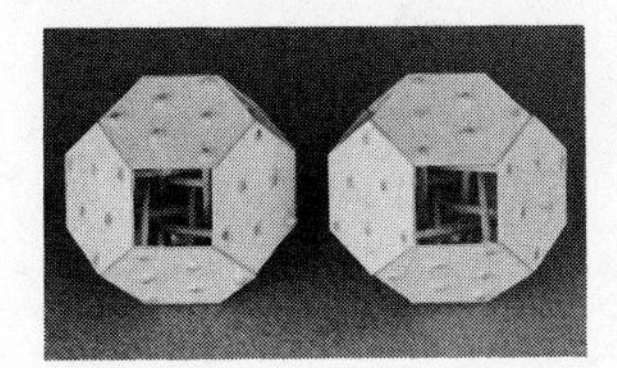

alan holden

with photographs by
doug kendall

columbia university press
new york and london

Alan Holden is the author of *The Nature of Solids.* In 1968 he received the Robert Andrews Millikan Award from the American Association of Physics Teachers for his contributions to the teaching of physics and related subjects.

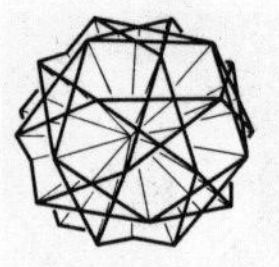


International Standard Book Number:
0-231-03549-7
Library of Congress Catalog Card
Number: 71-158459
Printed in the United States of America
10 9 8 7 6 5 4 3 2

**Preface.** Space provides no three-dimensional blackboard. We learn about space only by living in it. A child climbing in his jungle-gym may learn more about it than he will ever learn again, for his books will be made of two-dimensional sheets of paper.

Plato gave stature to the visualization of space by honoring **stereometry** as one of the four primary branches of his mathematics, cognate with arithmetic, geometry, and astronomy. Men of learning in ancient Greece took especial concern for "the putting together of the cosmic figures," their regular polyhedra, whose mathematical elegance and beauty inspire inevitable wonder. Today the heritage of that wonder comprises still more mathematical solids, discovered through centuries of thoughtful looking. They receive practical application by engineers and builders, idealized embodiment by chemists and crystallographers, imaginative extension by architects and sculptors.

The best way to learn about these objects is to make them, next best to handle them. Printed pictures are poor substitutes for moving pictures but better than words alone. This book can offer only pictures and a few words to connect them. It closes with its most important words, describing a way to make these beautiful shapes.

Alan Holden

# shapes, space, and symmetry

They are renowned as **the Platonic solids**—the tetrahedron, cube, octahedron, dodecahedron, icosahedron. Held in the hands and turned slowly about, their simple beauty seems ineffable. A geometer would account for it thus: the faces on each are regular polygons of a single species; and on each all the corners are alike.

Probably no one person first thought of these famous objects. The cube seems the easiest to think of, the dodecahedron perhaps the hardest. But there is evidence that the Etruscans used pentagonal dodecahedra as dice early in the first millenium B.C. In any case, by the fourth century B.C. the five solids have all appeared at the hands of Plato's close friend Theaetetus. Plato himself described them in his dialogue *Timeaus.* And two centuries later Euclid crowned his great work, the *Elements,* with a discussion of their geometry.

Can there really be *no more than five?* Begin with the equilateral triangle, the simplest regular polygon. To make a solid, at least three must join at any corner. Four can join at a corner, and so can five. But when six join at a corner they lie flat, to make a regular hexagon. Turn to the square: three can join at a corner, but four joining lie flat. Three regular pentagons can join at a corner, but more will not fit together. Three regular hexagons joining at a corner lie flat; three regular polygons with more sides cannot join at a corner. You have finished.

Can there really be *as many as five?* It is a little astonishing: by adding more polygons of the same kind, the boxes can be closed in every case. Four, eight, and twenty triangles make the tetrahedron, octahedron, and icosahedron. Six squares make the cube. Twelve pentagons make the dodecahedron. In each case no crevices remain.

If you are not astonished by that success, digress for a moment to examine the **convex deltahedra:** there are only eight, and there "should be" nine. They are the convex polyhedra that can be made of equilateral triangles, and they are called *deltahedra* because an equilateral triangle looks like the Greek capital letter *delta.*

The argument that there should be nine goes like this. Each triangle has three sides, and therefore the triangles that make a deltahedron with N faces have 3N sides altogether. When the triangles join to make a deltahedron, each side joins a side of another triangle to make an edge, and therefore the deltahedron has 3N/2 edges. Since "half an edge" has no meaning, N must be a multiple of two. Since at least three triangles must meet at any corner, to give solidity, and at most five triangles can meet at any corner, to retain convexity, the tetrahedron is the smallest convex deltahedron and the icosahedron is the largest. And there should be seven more, with six, eight, ten, twelve, fourteen, sixteen, and eighteen faces. But in fact the convex deltahedron with eighteen faces cannot be made!

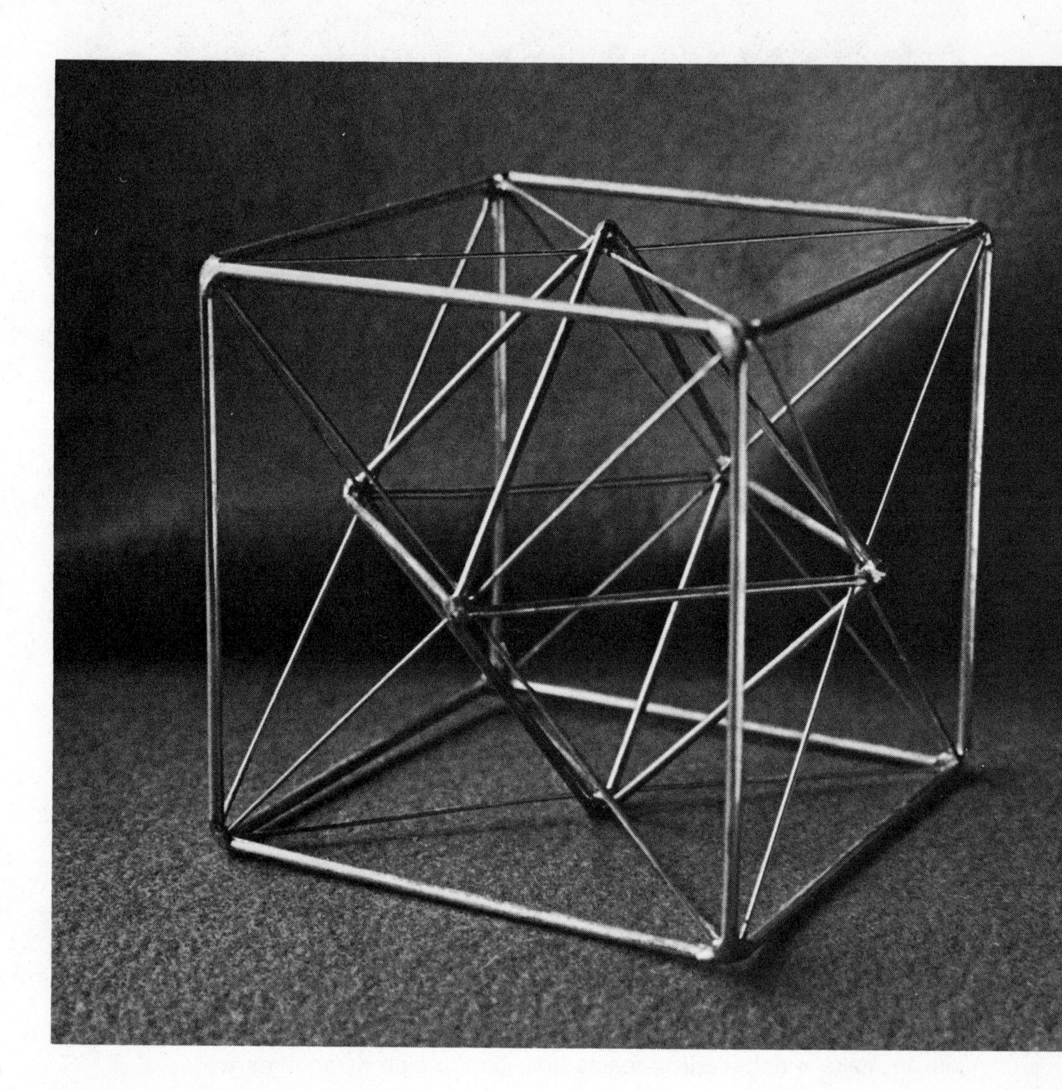

The Platonic solids may seem at first to be wholly unrelated to one another, except by the fact that each is made of regular polygons. But look now at a relation, called **duality,** that pairs them off with one another. Marking a point in the middle of each face of a cube, and connecting those points by lines, you draw the edges of an octahedron. And doing the same thing to an octahedron, you outline a cube. The cube and the octahedron are said to be *dual* to each other.

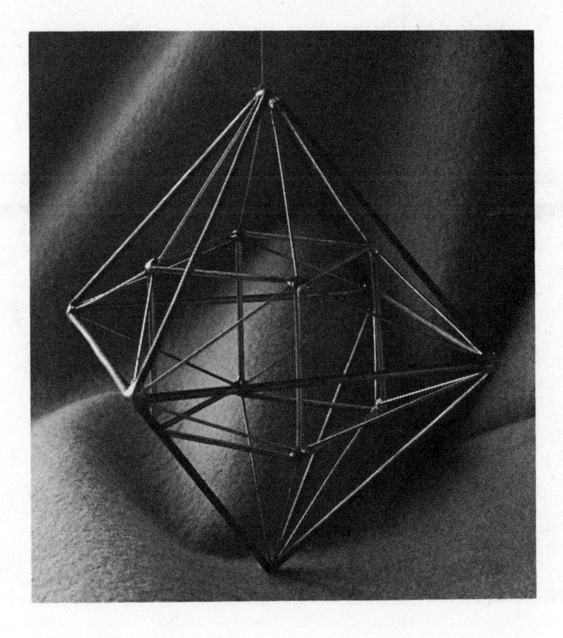

Notice that this relation does not depend on the fact that these figures are made of regular polygons. If the cube were distorted by squashing it, the cube would still be dual to a new sort of squashed octahedron. Their duality depends only on the fact that the cube has six four-sided faces, meeting by threes in eight corners, and the octahedron has eight three-sided faces, meeting by fours in six corners.

In the same way, points in the center of the faces of a pentagonal dodecahedron can be connected to outline the faces of an icosahedron. Conversely, points on the faces of an icosahedron form the corners of a dodecahedron. Here again is a dual pair. The operation demonstrating duality can be viewed as the replacement of faces by corners and of corners by faces: in other words, the operation interchanges corners and faces. The twelve faces and twenty corners of a dodecahedron become the twelve corners and twenty faces of an icosahedron. It may seem most natural to think of these solids as made of their faces; but it is just as valid to think of them as made of their corners.

Turning at last to the tetrahedron, you find that the interchange of corners and faces produces another tetrahedron. To be sure the new one is turned in a different direction from its parent. But the two are geometrically similar: a tetrahedron is *self-dual.*

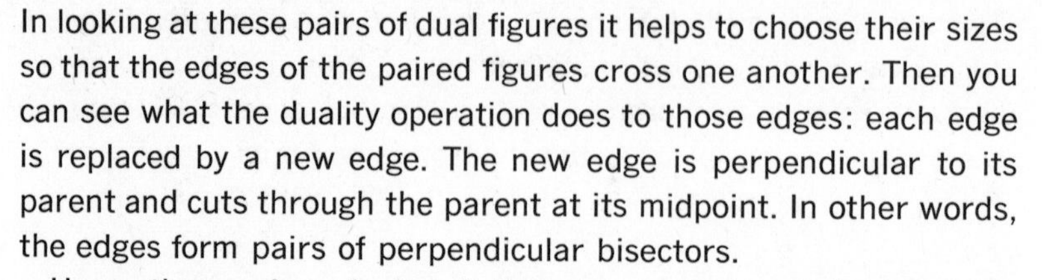

In looking at these pairs of dual figures it helps to choose their sizes so that the edges of the paired figures cross one another. Then you can see what the duality operation does to those edges: each edge is replaced by a new edge. The new edge is perpendicular to its parent and cuts through the parent at its midpoint. In other words, the edges form pairs of perpendicular bisectors.

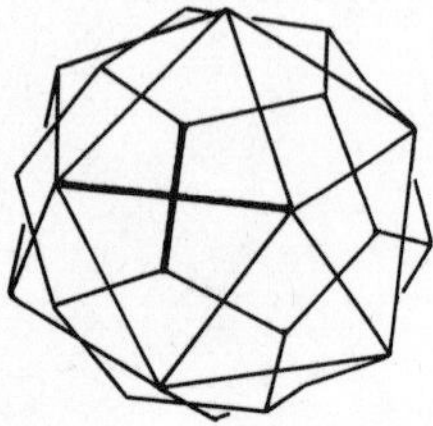

Hence the number of edges is the same on each member of a dual pair; a cube and an octahedron both have twelve edges, and a dodecahedron and an icosahedron have thirty. The table summarizes all this.

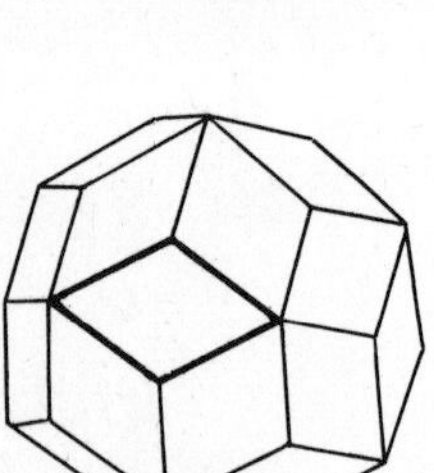

| *Platonic solids* | *faces* | *corners* | *edges* |
|---|---|---|---|
| tetrahedron | 4 | 4 | 6 |
| cube | 6 | 8 | 12 |
| octahedron | 8 | 6 | 12 |
| dodecahedron | 12 | 20 | 30 |
| icosahedron | 20 | 12 | 30 |

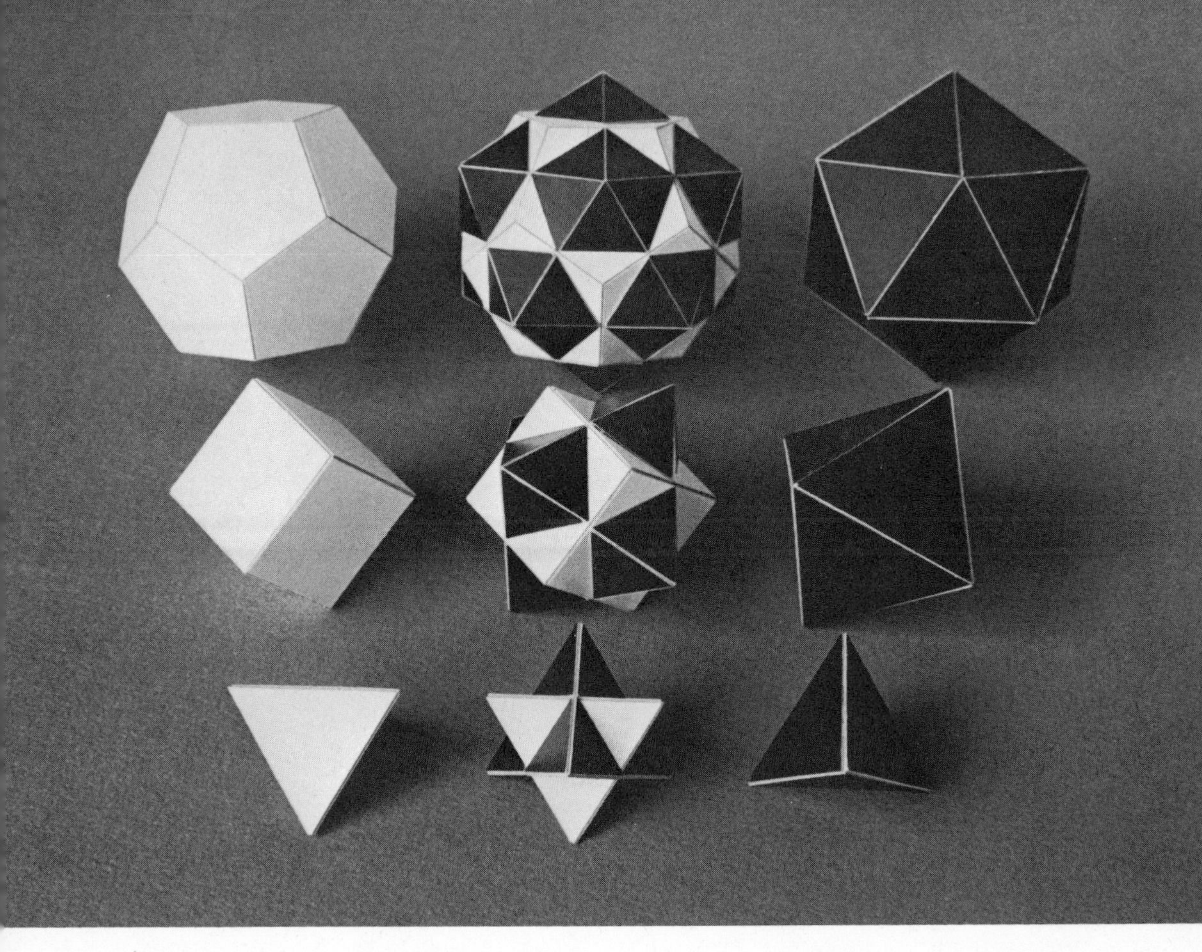

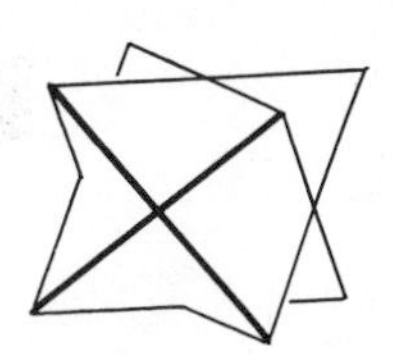

An elegant way of showing the dualities among these solids is to make cardboard models in which the two members of a dual pair appear to *interpenetrate* each other. The easiest strategy is first to make the three white solids, and then to attach black pyramids to their faces. Since the faces of the black solids are equilateral triangles, the faces of the projecting pyramids are also equilateral triangles. Notice that each pair of dual edges forms the diagonals of a rhombus whose vertices are the corners of the model. For the model of two tetrahedra, that rhombus is a square.

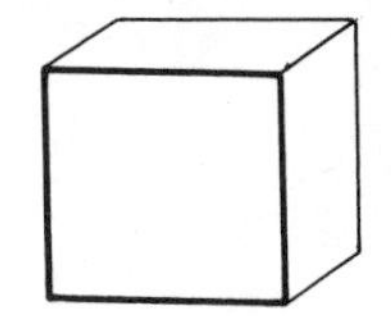

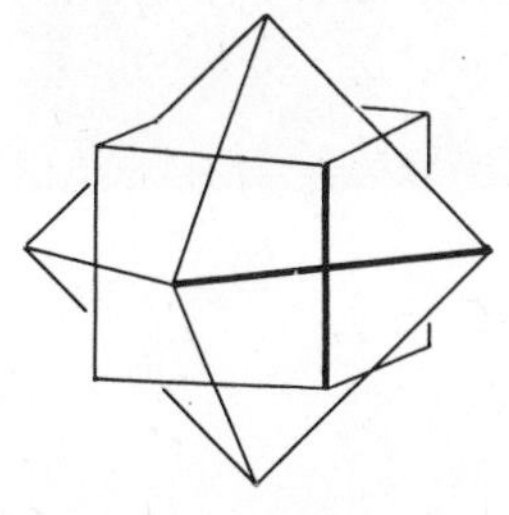

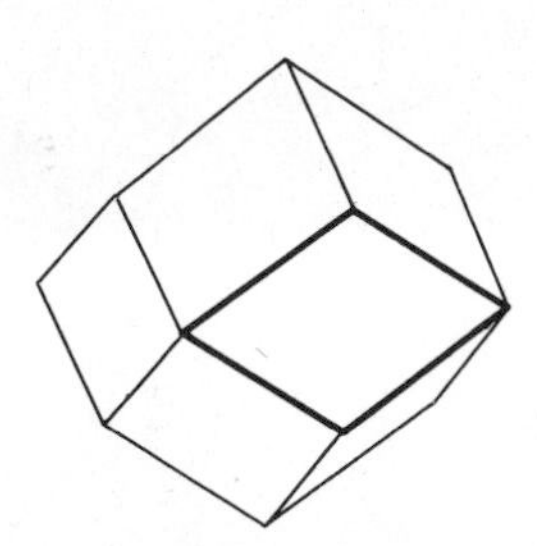

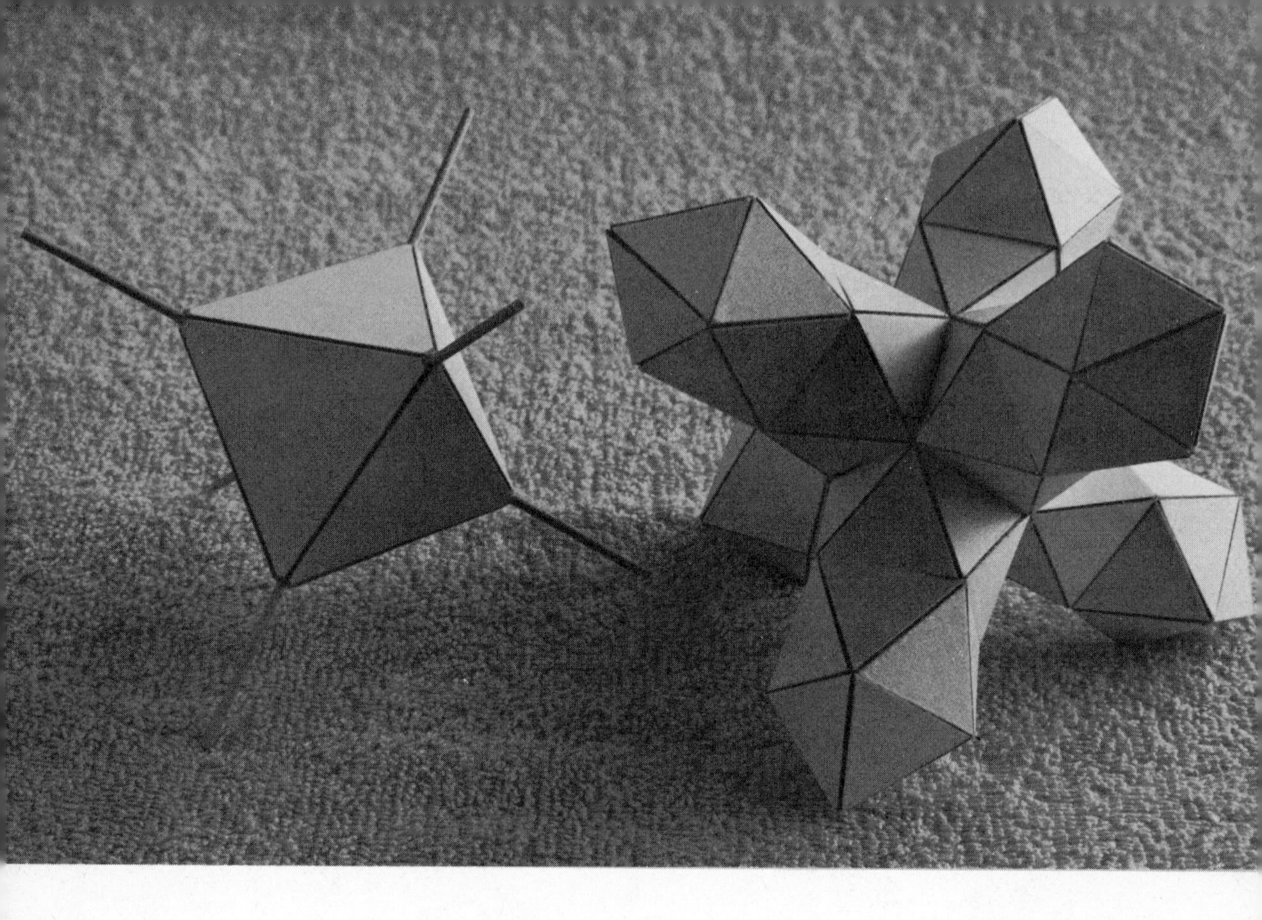

Upon the corners of an octahedron, or the faces of a cube, you can build six four-sided towers, protruding in three pairs of opposites.

Upon the faces of an octahedron, or the corners of a cube, you can build eight three-sided towers, protruding in four pairs of opposites.

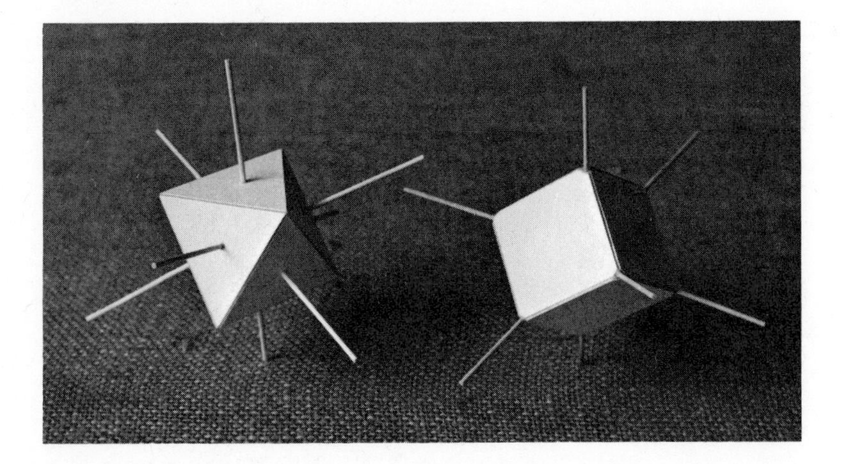

Upon the edges also you can build towers, twelve protruding in six pairs of opposites. The directions in which the towers protrude, marked by the sticks in these models, are **axes of rotational symmetry.** Turning the cube or the octahedron about one of these sticks by half of a full revolution, you find that the whole figure looks just as it would if you had not turned it. Such a direction is called an *axis of twofold rotational symmetry* because the figure looks the same twice in one revolution, first after half a revolution and again after the full revolution. Similarly, the directions shown on the preceding pages are *axes of fourfold symmetry* and *axes of threefold symmetry.*

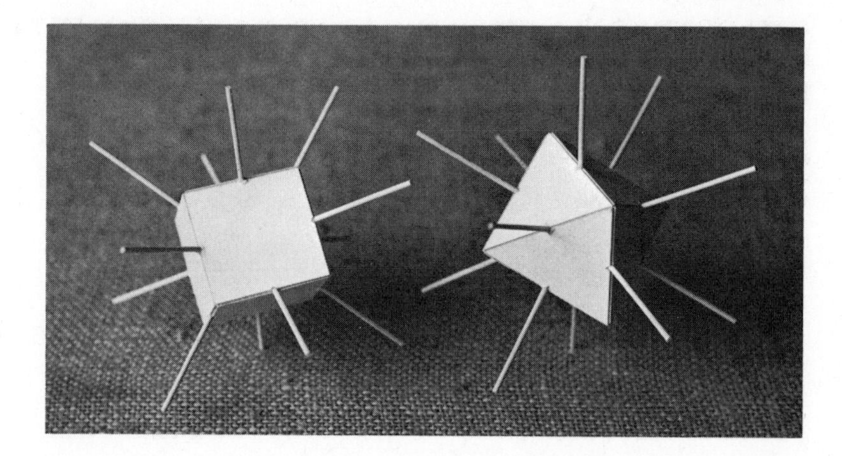

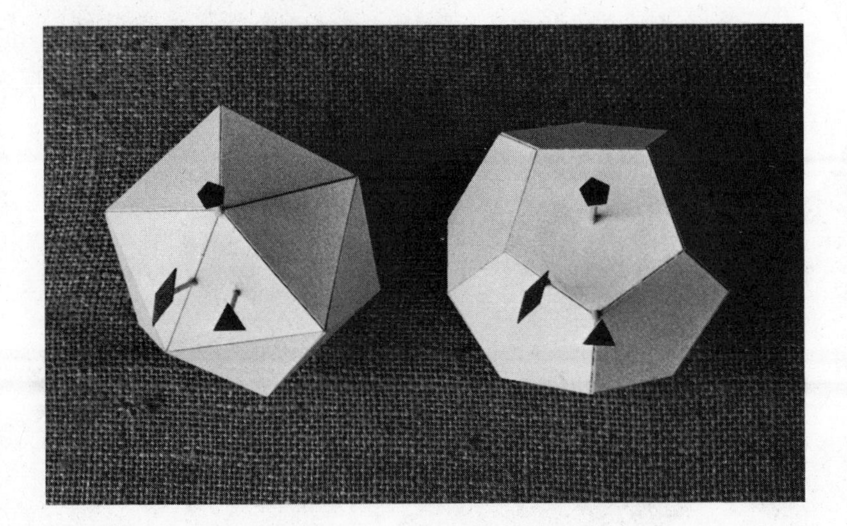

A corresponding examination of the dodecahedron and the icosahedron reveals axes of fivefold, of threefold, and of twofold symmetry. For the dodecahedron the fivefold axes pass through pairs of opposite faces, the threefold axes through pairs of opposite corners, and the twofold axes through pairs of opposite edges. The duality of the icosahedron with the dodecahedron then tells where the axes must lie in the icosahedron. And the earlier count of faces, corners, and edges tells how many of these axes there are for each figure: six fivefold, ten threefold, and fifteen twofold axes. Since a model showing all these axes would bristle with sticks, the models pictured above show only one typical axis of each sort.

You can now easily see for yourself that the regular tetrahedron has only four threefold and three twofold axes of symmetry. Hence the symmetry axes of the Platonic solids are:

| *Platonic solids* | *twofold* | *threefold* | *fourfold* | *fivefold* |
|---|---|---|---|---|
| tetrahedron | 3 | 4 | — | — |
| cube<br>octahedron | 6 | 4 | 3 | — |
| dodecahedron<br>icosahedron | 15 | 10 | — | 6 |

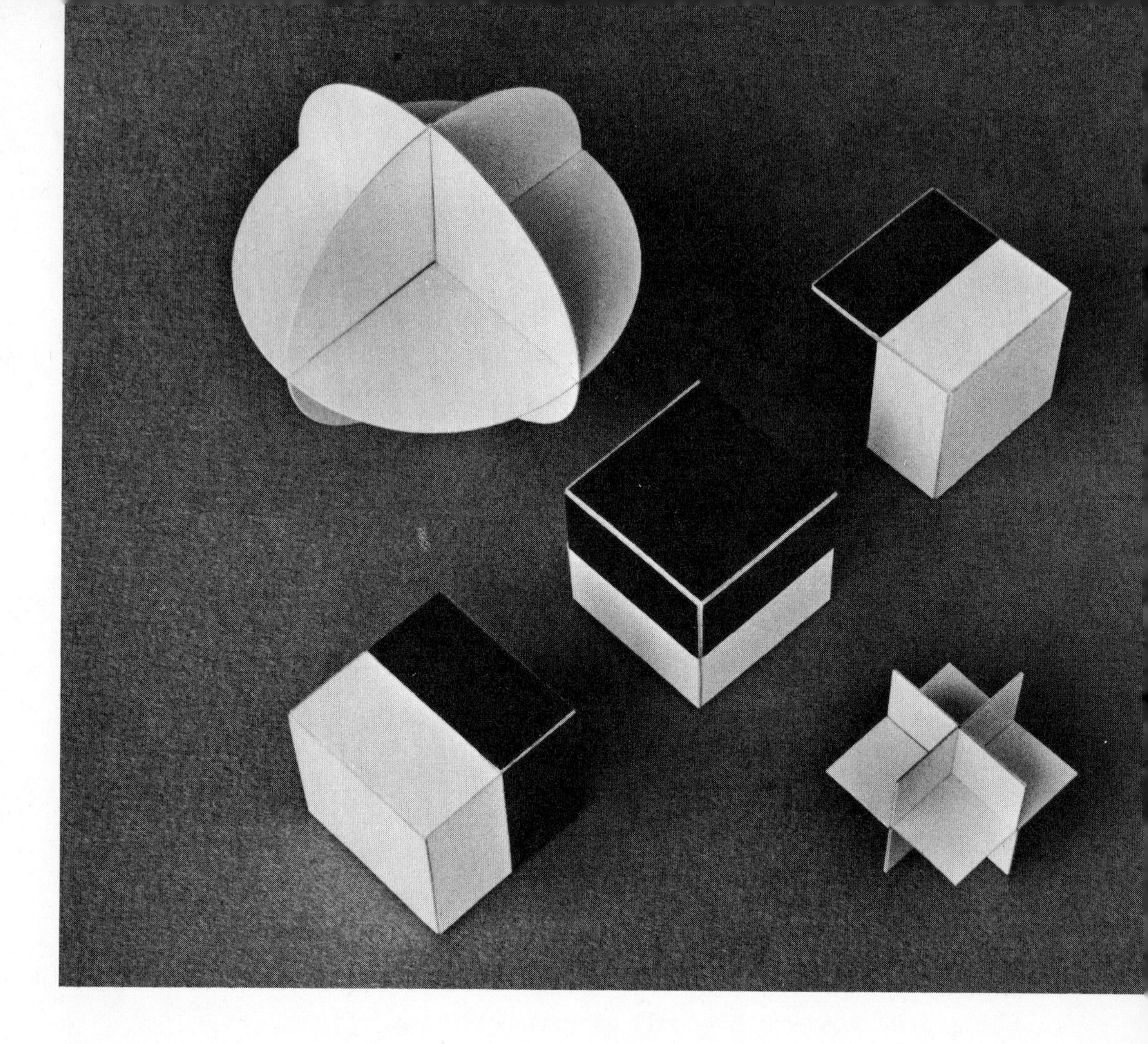

The symmetries of the Platonic solids are not completely described by their axes of rotational symmetry; they have also **planes of reflection symmetry.** A flat surface that cuts the cube in half parallel to a pair of opposite faces marks such a plane. If the half-cube were placed upon a mirror, the reflection in the mirror would build upon the half-cube another half-cube, constructing the appearance of an entire cube. Clearly there are three such planes, parallel to the three pairs of opposite faces. The models show *(lower right)* how the faces of the cube cut off those planes, and *(upper left)* how an enclosing sphere would cut them off.

By now it will not surprise you that the figure dual to the cube, the regular octahedron, also has three planes of reflection symmetry. Comparing the way the faces of the octahedron cut off those planes *(lower right)* and the way a sphere cuts them off *(upper left)* makes clear that an octahedron fits neatly into a sphere. That is to say, if an octahedron were surrounded by a big spherical soap bubble, and the air were slowly let out of the bubble, the contracting bubble could touch all corners of the octahedron at the same time. A geometer would say that an octahedron can be *inscribed* in a sphere touching its corners.

A cube has other planes of reflection symmetry dividing it in half. A reflection plane passes through each pair of opposite edges; and since the cube has six such pairs of edges, it has six such planes. Here the comparison of how the cube cuts off those planes and how a sphere cuts them off shows that a cube, like an octahedron, can be inscribed in a sphere that touches its eight corners.

Of course an octahedron has six similar planes of symmetry. Altogether, therefore, a cube or an octahedron has nine planes of reflection symmetry. Three of those planes are perpendicular to its three axes of fourfold rotational symmetry, and six are perpendicular to its six axes of twofold rotational symmetry. All the planes pass through one common point, the center of the cube or the octahedron.

The dodecahedron and icosahedron have planes of reflection symmetry also; each passes through a pair of opposite edges and cuts another pair of opposite edges. Each is perpendicular to an axis of twofold rotational symmetry, and therefore there are fifteen of those planes. Looking at how a sphere cuts them off, you see that they meet in threes at the corners of a dodecahedron, and in fives at the corners of an icosahedron. Clearly those two figures can also be inscribed in a sphere.

The tetrahedron, in addition to its four threefold axes and three twofold axes of rotational symmetry, has six planes of reflection symmetry, each containing one of its six edges. Those planes are arranged in the same way as six of the reflection planes of the cube and the octahedron. By inscribing a tetrahedron in a cube, you can see why the arrangements of those planes correspond.

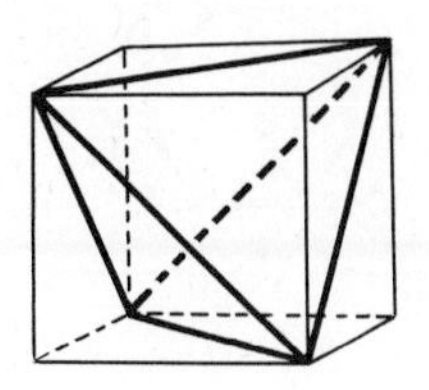

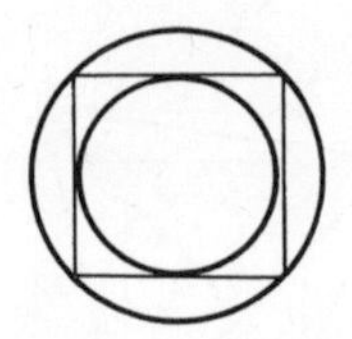

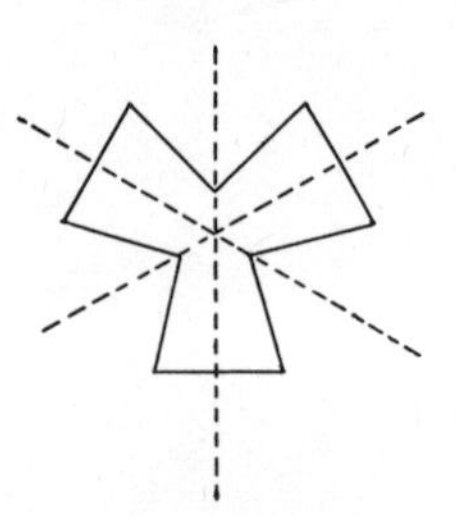

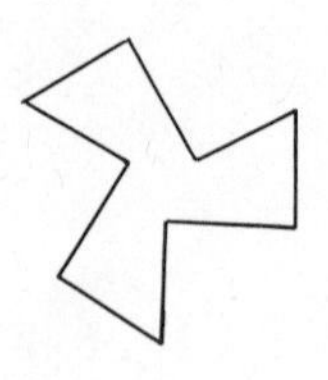

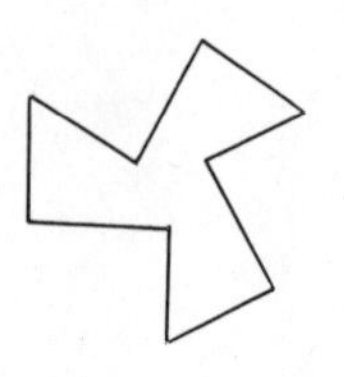

The ideas of *duality* and of *symmetry* are powerful intellectual instruments, often providing quick routes to conclusions that may otherwise seem hard to reach. Think again of the observation that each of the Platonic solids can be inscribed in a sphere. Then recall that each of them can also be inscribed in its dual, in such a way that its corners fall at the centers of its dual's faces. Then the sphere that inscribes it will touch the center of each of those dual faces. But the dual solid is also a Platonic solid. In other words, a spherical soap bubble, expanding inside a Platonic solid, can touch the centers of all its faces at the same time. Thus any of these solids can not only be *inscribed in* a sphere but can also be *circumscribed around* a smaller sphere, just as a square can define both a circumscribed circle and an inscribed circle.

Turn next to look at one remarkable aspect of the idea of geometrical symmetry. In searching for axes of rotational symmetry, you could pick up a solid and turn it about some direction through its center. In searching for planes of reflection symmetry, however, you had to use your imagination, or perhaps some optical trick such as a mirror. This distinction between *performable* operations such as a rotation and *nonperformable* operations such as a reflection has a consequence illustrated by the figures at the left. Each figure has an axis of threefold rotational symmetry, and the uppermost has also three planes of reflection symmetry. The other two figures do not have such planes, and therefore, though they look much alike, they cannot be turned so that they coincide with each other. They form a pair like a right hand and a left hand in which each is the reflection of the other. Such a "right-left" distinction is never possible when an object has a nonperformable symmetry; it is always possible when it lacks that sort of symmetry.

Return now to planes that cut the Platonic solids in half, and examine some interesting planes that are not planes of reflection symmetry. The three planes perpendicular to the tetrahedron's three twofold symmetry axes correspond with three symmetry planes of the cube and the octahedron, and here as before each cuts the solid to yield a square cross-section. But they are not symmetry planes of the tetrahedron because the half-tetrahedron and its reflection do not reproduce the tetrahedron: they do so only if the two halves are turned with respect to each other by a quarter of a full turn.

A plane perpendicular to a threefold rotation axis of an octahedron or a cube cuts the solid in a regular hexagonal cross-section. Since there are four such axes, there are four possible hexagonal sections.

Each plane perpendicular to one of the six fivefold rotation axes of a dodecahedron or an icosahedron cuts the figure to give a regular decagon as cross-section. Notice that none of these planes is a true

plane of reflection. In each case the reflection can be made to coincide with the discarded half only by turning it through half the angle characterizing the rotation axis perpendicular to the plane.

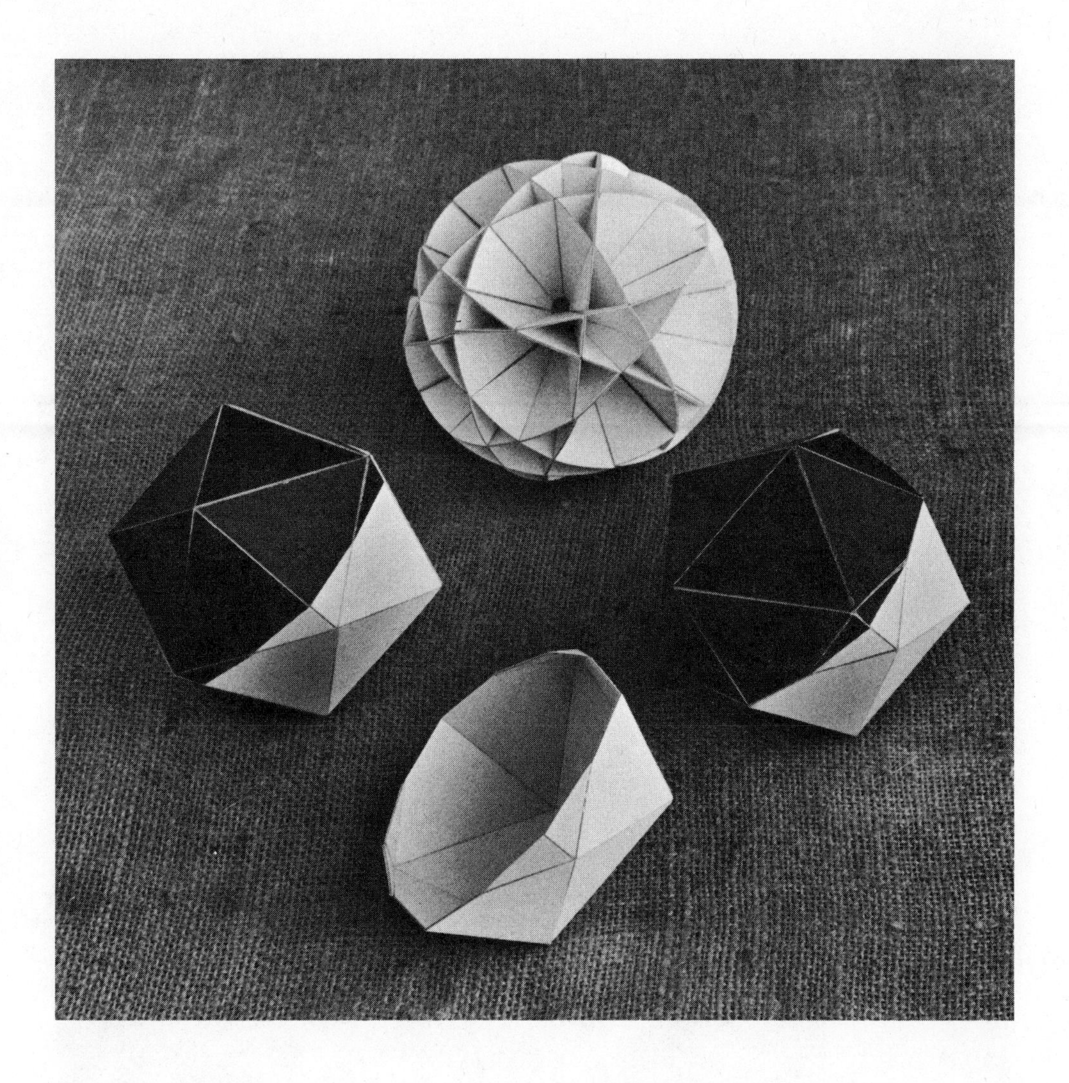

The ten planes perpendicular to the threefold rotation axes of a dodecahedron cut the figure in regular hexagons, just as the four corresponding planes cut the cube and the octahedron.

But the icosahedron is exceptional: its cross-sections perpendicular to the threefold axes are not regular polygons. Their shape is a dodecagon that has threefold symmetry and whose sides all have the same length, but whose angles are not all the same.

When halves of the Platonic solids having the same cross-sections come together, monsters are born. Often the two halves can be joined across their common cross-section in two different ways. When half a dodecahedron meets half an icosahedron on a decagonal cross-section, the monster looks like a dodecahedron when it is viewed in one direction along its fivefold rotation axis, and like an icosahedron in the opposite direction. A cube and an octahedron can join at a hexagonal cross-section, and either of these figures could join with a dodecahedron.

Further insight into the Platonic solids comes from looking at ways of inscribing them in one another. You saw that a systematic way to inscribe a solid in its dual is to place its corners in the middle of the dual's faces. Then all the rotation axes and reflection planes of the two solids coincide. That fact suggests a systematic way to inscribe solids in one another when they are not duals. For example, a tetrahedron can be placed in a cube so that the four threefold axes of the figures coincide, and then it can be expanded until the corners of the tetrahedron fall on four of the cube's eight corners. The duality of cube and octahedron suggests how to inscribe the tetrahedron correspondingly in an octahedron.

A tetrahedron can be inscribed in a dodecahedron by following the same principle: the four threefold axes of the tetrahedron can be aligned with four of the ten threefold axes of the dodecahedron. It turns out that then the three twofold axes of the tetrahedron are automatically aligned with three of the twofold axes of the dodecahedron. Again a tetrahedron in this position can be expanded until its corners coincide with four corners of the dodecahedron. In this case the edges of the tetrahedron do not lie on faces of the dodecahedron, as they lie on faces of its inscribing cube.

There is, however, a more important difference between a tetrahedron in a cube and a tetrahedron in a dodecahedron. The six reflection planes of an inscribed tetrahedron coincide with six of the cube's reflection planes, but they do not coincide with any of the reflection planes of the inscribing dodecahedron. Hence the combi-

nation of tetrahedron and dodecahedron has no surviving reflection planes. Since the combination lacks nonperformable symmetries, it can appear in two essentially different forms, right-handed and left-handed, each the mirror image of the other.

Starting from this inscription in a dodecahedron, you can use the principle of duality for inscribing a tetrahedron in an icosahedron. Each corner of the tetrahedron falls in the middle of a face of the icosahedron, and therefore the opposite face of the icosahedron is parallel to a face of the tetrahedron. You can think of the tetrahedron as picking out for special notice four faces of the icosahedron, either the four that carry the tetrahedron's corners or the four that are parallel to the tetrahedron's faces. The twenty faces of an icosahedron could be sorted into five groups of four, with each group "belonging to" one or another tetrahedron.

When a cube is inscribed in an icosahedron by aligning the three-fold axes, the corners of the cube fall on faces parallel to those of a regular octahedron. But it is more interesting to notice that each *face* of the cube is parallel to an *edge* of its inscribing icosahedron or dodecahedron. Since each of those inscribing figures has thirty edges, they can be sorted into five groups of six edges so that each group belongs with one or another cube. Here three reflection planes of the inscribed cube coincide with three of the inscribing figure's reflection planes. The combination of the two figures therefore retains those planes of reflection, and there is no distinction between right-handed and left-handed forms of the combination.

The correlation of six edges of a dodecahedron with the six faces of a cube brings to notice a way of building a dodecahedron upon an inscribed cube by attaching six identical "roofs" to the cube's faces. The ridgepole of each roof forms one of the edges of the dodecahedron, and each face of the dodecahedron is formed of two parts of adjacent roofs that line up with each other.

When two cubes of the same size put their centers in the same place, they can do so in an infinite variety of ways, interpenetrating each other so that parts of each cube are hidden by its partner and other parts are exposed. In the most interesting of these many ways, the two cubes share one or another of their rotation axes. The two cubes at the left have one of their threefold axes of rotation in common; each is turned about it 60 degrees with respect to the other. The cubes at the right have a common twofold axis, and are turned with respect to each other by 90 degrees about it.

In the first of these **compound solids** the two cubes share two of their corners, but in the second the corners are all separated. It is tempting to think of the second as a new regular solid with twelve identical square faces that intersect one another, and with sixteen corners, each identically surrounded by three faces. But the solid is not called a new regular solid because it can be decomposed into two interpenetrating cubes.

A compound solid looking even more "regular" arises when a third cube joins the first two so that each shares two of its twofold axes, one with each of its partners. An unexpected symmetry appears: the compound has the high symmetry of the cube itself, three fourfold axes, four threefold axes, six twofold axes, and nine reflection planes. Most of the elements of symmetry do not coincide with those of the constituent cubes, but each of the three fourfold axes lies along a fourfold axis of one or another of the three cubes.

Taking that last fact as motivation, you can put four cubes together in such a way that each of the four threefold axes of the compound falls along the threefold axis of one or another of the constituent cubes. Again the product has the symmetry of a cube, though most elements do not coincide with those of the constituents. The compound at the left has the additional property that all its corners are identically surrounded by other corners, a property lacked by the compound at the right.

Combining cubes into very symmetrical compounds can be carried a step further by recalling the way of inscribing a cube in a dodecahedron. Each edge of the dodecahedron can be matched against a face of an inscribed cube, and five such cubes will account for the dodecahedron's thirty edges. The forty corners of the five inscribed cubes fall at the twenty corners of the dodecahedron, with two cube corners sharing a dodecahedron corner. Because of its method of

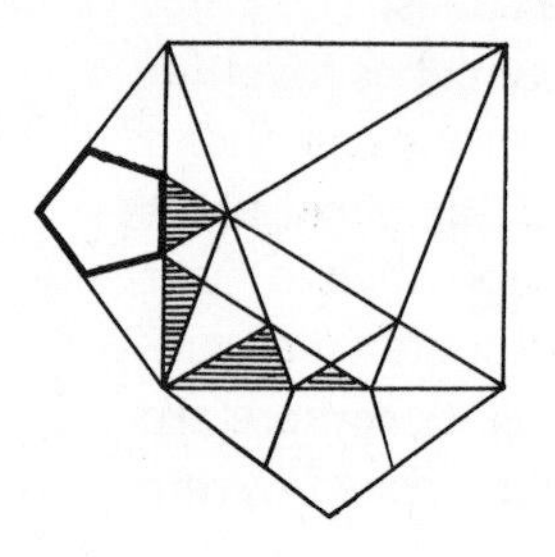

composition, the compound has the full symmetry of the dodecahedron, six fivefold axes, ten threefold axes, fifteen twofold axes, and fifteen reflection planes. Notice that here, with the appearance of fivefold axes, a kind of symmetry arises in the compound that is not possessed by the constituent cubes. The diagramed construction of the exposed parts on the model begins by extending three sides of a regular pentagon.

With equal drama, an unexpected symmetry appears when two dodecahedra are compounded so that they are turned 90 degrees with respect to each other about a common axis of twofold symmetry. The compound has the full symmetry of a cube. All fivefold symmetries have disappeared, and three fourfold axes have arisen that have no counterparts in a dodecahedron.

The inscription of a tetrahedron in a dodecahedron suggests a way of compounding five tetrahedra in a symmetrical fashion. After a single tetrahedron has been inscribed, four more appear when the assembly is turned about a fivefold rotation axis of the dodecahedron. The five-times-four corners of the tetrahedra occupy the twenty corners of the dodecahedron.

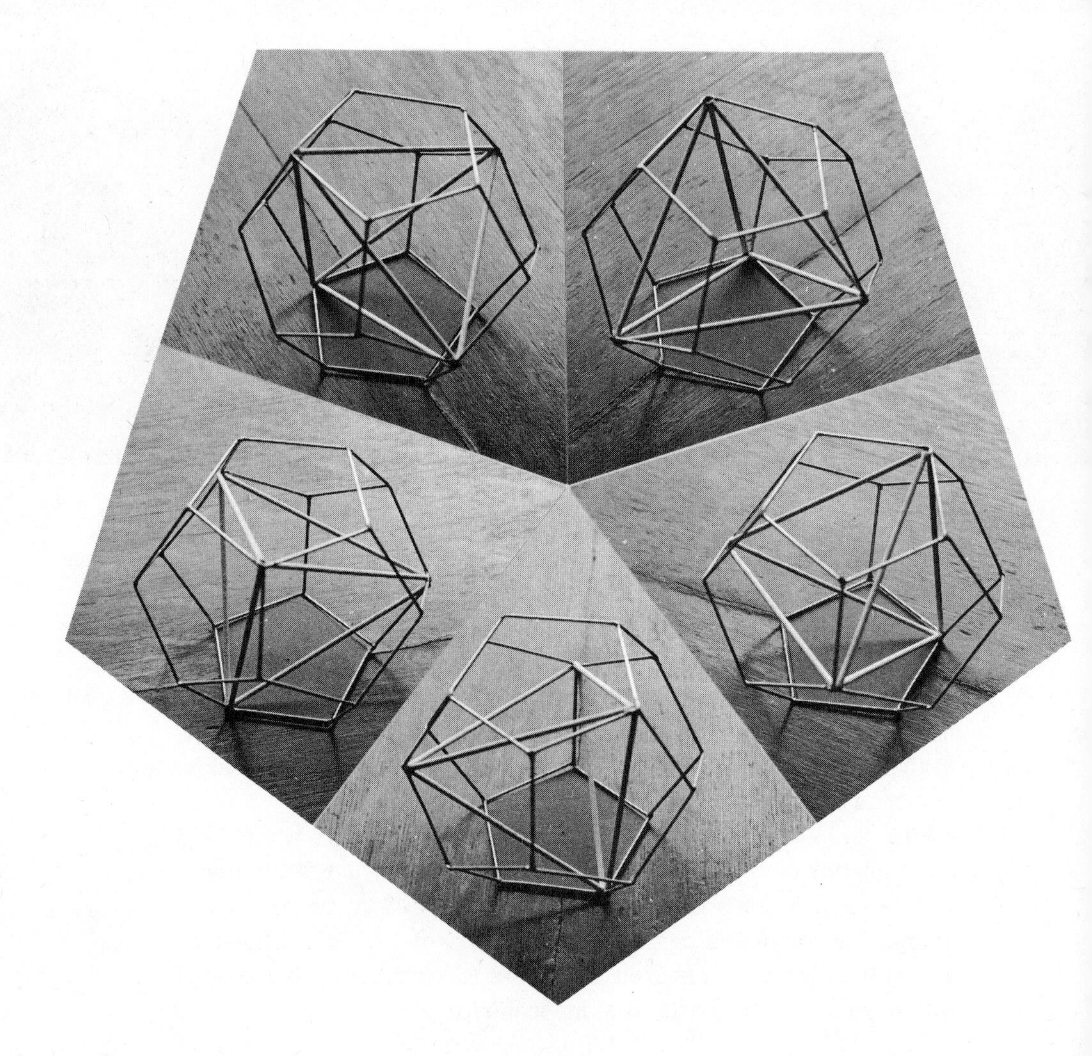

Remember that there are two different ways to inscribe a tetrahedron in a dodecahedron, providing a left-handed and a right-handed inscription. Turning the assembly therefore generates either a left-handed or a right-handed compound. All the exposed parts of the faces in each of these compounds have the same shape, and the shapes for the two compounds are mirror images of each other. Each of the compounds has all the rotation axes of a dodecahedron, but none of its reflection planes.

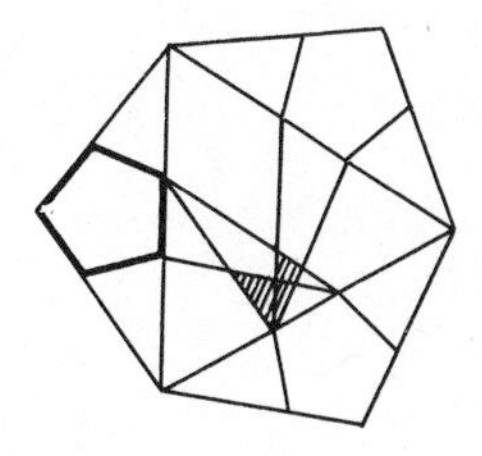

When you chop off the tips of the big black cube at the left, you replace its eight corners with little white equilateral triangles. The chopping is called **truncation.** Chopping further, you reach a stage at which the triangles meet at their vertices; still further, the triangles become hexagons. Meanwhile the relics of the original faces on the cube become smaller and finally disappear, leaving a regular octahedron. Starting again, with the big black octahedron, truncation yields the same succession of solids in reverse order and ends with a cube. Notice a special property of the stages of truncation shown above. At each stage all the faces of the solid are regular polygons—equilateral triangles, squares, regular hexagons, or regular octagons.

These solids therefore share with the Platonic solids the property of having only regular polygons for their faces. They differ from the Platonic solids in having two sorts of polygons for their faces instead of just one sort. But they share another property with the Platonic solids: all the corners on each are alike. A corner can be described by first noticing what polygons meet at the corner and then traveling around the corner, naming those polygons in sequence. The sequence in the first truncation of the cube is octagon-triangle-octagon: all its corners yield the same sequence.

Solids whose faces are made of several sorts of regular polygons and whose corners are all alike are called **semiregular solids.** The term "regular solids" is reserved for those, such as the Platonic solids, whose faces are all made of regular polygons of a single sort. Some of the compounds examined earlier are therefore "regular compounds," but some others are not because their corners are not all alike.

The semiregular solids share still another property with the Platonic solids. Each can be inscribed in a sphere that touches all its corners. Since they have more corners than the Platonic solids, they provide closer approximations to spheres. Their spherical appearance becomes conspicuous in the semiregular solids generated by truncating the icosahedron or the dodecahedron.

It seems natural to associate the truncations of the cube and the dodecahedron that yield solids on which triangular faces meet at their vertices. Each of them stands half way between the extremes of truncation. For this reason they are named the *cuboctahedron* and *icosidodecahedron.* Notice that the number of faces of these solids is the sum of the numbers of faces on the cube and octahedron, and on the dodecahedron and icosahedron, and the faces have the same shapes as the faces on those solids. The number of corners is easily counted by noticing that truncation has obliterated the original corners and has placed a new corner in the middle of each of the original edges: the cuboctahedron has twelve corners and the icosidodecahedron thirty.

Truncating the tetrahedron produces a semiregular solid with four triangular and four hexagonal faces, and with twelve corners. Clearly further truncation, carried to the point where the vertices of the triangles meet, would yield an octahedron and not a new semiregular solid. Here is another example of the unique behavior of the tetrahedron, closely related to the fact that it is self-dual.

Again it appears natural to associate two semiregular solids, the *truncated octahedron* and the *truncated icosahedron,* where truncation has converted the parental triangles into regular hexagons. On each the number of faces is equal to the sum of the numbers of corners and of faces on its truncated parent. The number of corners is double the number of edges on the parent.

On the right above appear two new semiregular solids that did not occur in the course of truncating the Platonic solids. Here square faces replace the corners of the cuboctahedron and the icosidodecahedron, and it may seem at first that they could be obtained by truncating the corners of those solids. But closer examination shows that such a truncation will yield rectangles that are not square. The new solids are called the *small rhombicuboctahedron* and the *small rhombicosidodecahedron.* Notice in the latter case that the faces comprise three different species of regular polygons, not just two, but all corners are alike nevertheless.

On the right the *truncated cube* and the *truncated dodecahedron* seem naturally associated. The regular octagons of the first and the regular decagons of the second appear again in the *great rhombicub-octahedron* and the *great rhombicosidodecahedron* on the left, each made of three species of regular polygons, which cannot be produced by truncating the Platonic solids.

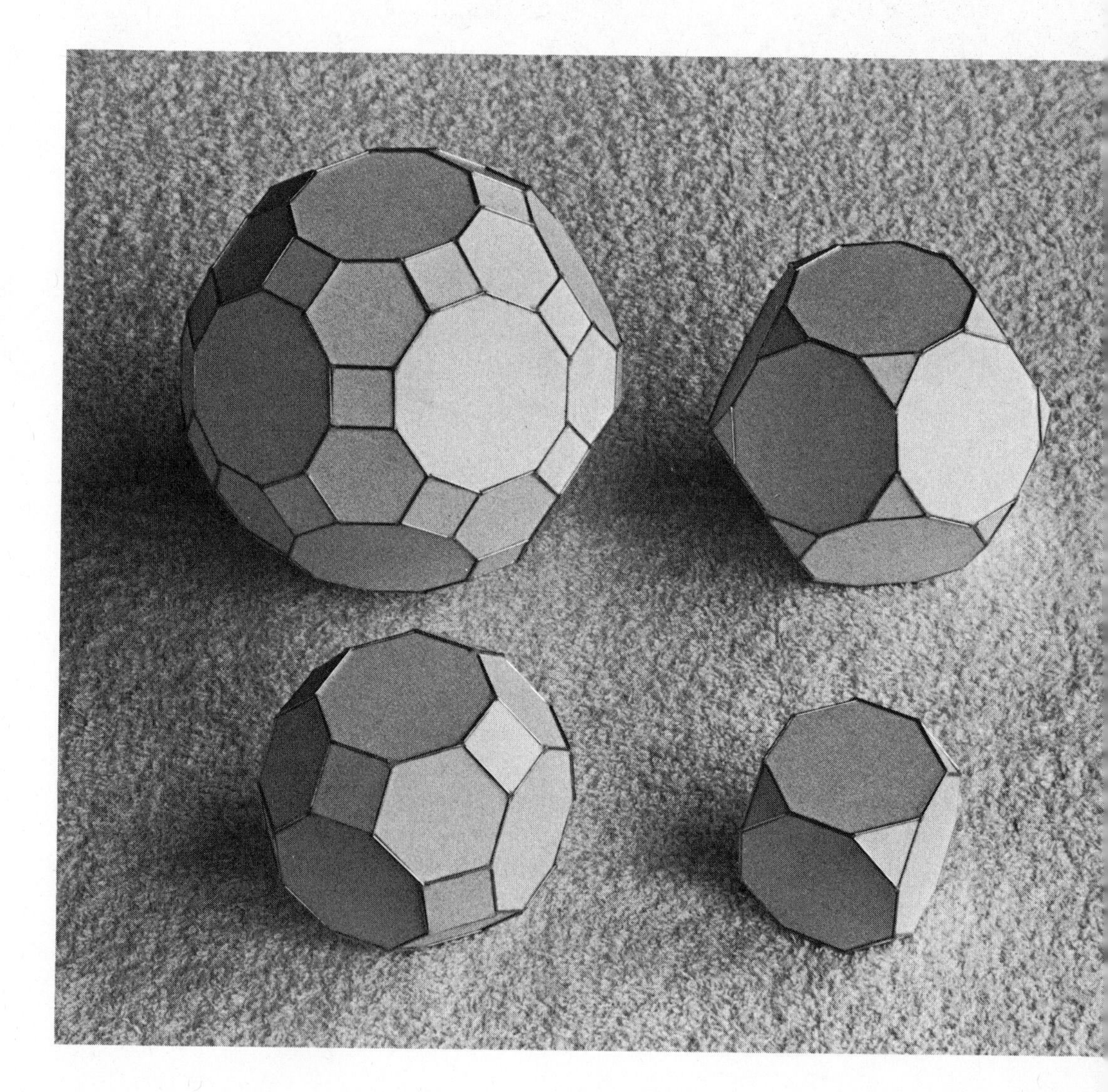

The most surprising of the semiregular solids are the *snub cube* and the *snub dodecahedron.* Usually counted as two, they really furnish four distinguishable solids. The snub cube has all the rotation axes of a cube, and the snub dodecahedron all the axes of a dodecahedron. But these solids have no planes of reflection, and therefore they appear in right-handed and left-handed forms.

Notice a simple way to count the corners on these solids. The snub dodecahedron has twelve pentagonal faces, and each vertex of a pentagon appears at a corner. Since only one pentagon appears at each corner, and since all corners are alike, there must be twelve-times-five corners.

Should we count them as thirteen or fifteen? In any case, they are the only convex semiregular solids that have Platonic symmetries. Early in the Christian era the Alexandrian mathematicians Heron and Pappus asserted that Archimedes wrote about all these solids. Even though his writing is lost, they are often called the **Archimedean solids.**

According to Heron, Archimedes said that "Plato also knew one of them, the figure with fourteen faces, of which there are two sorts, one made up out of eight triangles and six squares . . . and the other made up out of eight squares and six triangles, which seems more difficult." Surely Heron misunderstood what Archimedes said, for the other fourteen-faced figures above are the truncated cube and the truncated octahedron.

The table shows the numbers of corners, edges, and faces of each species, for the thirteen Archimedean solids.

| *name* | *c* | *e* | *f3* | *f4* | *f5* | *f6* | *f8* | *f10* |
|---|---|---|---|---|---|---|---|---|
| truncated tetrahedron | 12 | 18 | 4 | — | — | 4 | — | — |
| truncated cube | 24 | 36 | 8 | — | — | — | 6 | — |
| truncated octahedron | 24 | 36 | — | 6 | — | 8 | — | — |
| cuboctahedron | 12 | 24 | 8 | 6 | — | — | — | — |
| small rhombicuboctahedron | 24 | 48 | 8 | 18 | — | — | — | — |
| great rhombicuboctahedron | 48 | 72 | — | 12 | — | 8 | 6 | — |
| snub cube | 24 | 60 | 32 | 6 | — | — | — | — |
| truncated dodecahedron | 60 | 90 | 20 | — | — | — | — | 12 |
| truncated icosahedron | 60 | 90 | — | — | 12 | 20 | — | — |
| icosidodecahedron | 30 | 60 | 20 | — | 12 | — | — | — |
| small rhombicosidodecahedron | 60 | 120 | 20 | 30 | 12 | — | — | — |
| great rhombicosidodecahedron | 120 | 180 | — | 30 | — | 20 | — | 12 |
| snub dodecahedron | 60 | 150 | 80 | — | 12 | — | — | — |

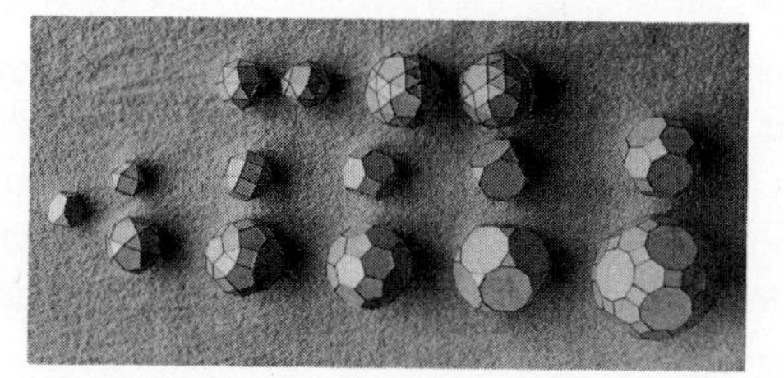

To fix firmly an image of the distinction between solids with and without nonperformable symmetries, contrast the figures outlined by wires. The figure above has all the symmetry of a cube, including its nine planes of reflection. On one of the figures below, the pyramids are all twisted in one sense, and on the other figure in the opposite sense. Each figure has all the cube's rotation axes but lacks its reflection planes.

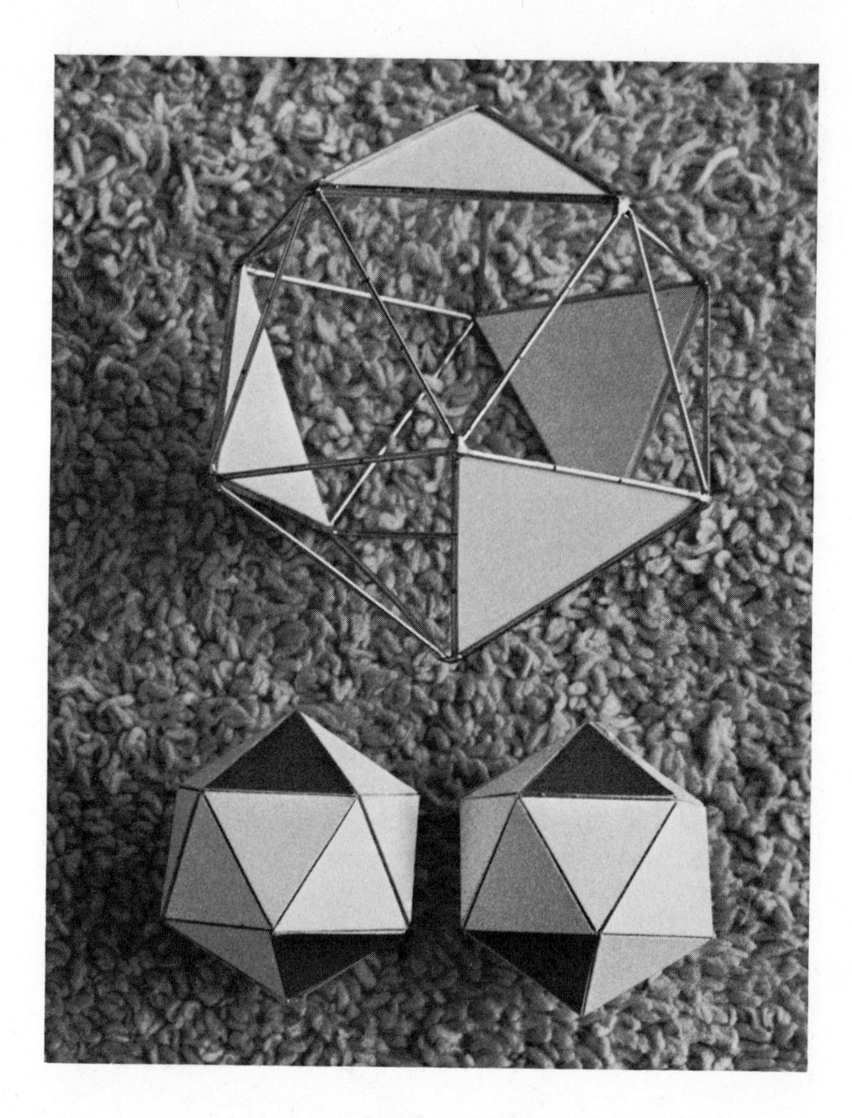

It is possible to make right-handed and left-handed choices of four faces on an icosahedron, in each of which the chosen faces are tetrahedrally disposed. The choice corresponds with choosing a left or right inscription of a tetrahedron in a dodecahedron, and hence with distinguishing the left and right regular compounds of five tetrahedra (see p. 39).

The combination of a cube inscribed in a dodecahedron retains three planes of reflection. But the planes are not necessarily retained throughout the course of building a dodecahedron by attaching "roofs" to a cube (see p. 33). Attaching only three of the six roofs produces a monster, half dodecahedron and half cube, and the monster can take either a right-handed or a left-handed form. Here the distinction becomes especially subtle. Viewed along an axis of threefold symmetry, either form of the monster looks like a dodecahedron from one direction and like a cube from the opposite direction, appearing to retain planes of symmetry in each case. You can apprehend the distinction only by handling the two monsters.

Unlike the Platonic solids, the Archimedean solids are not dual to one another. Replacing their corners by faces and their faces by corners produces new solids, the **Archimedean duals.** An especially well-known example, the white *rhombic dodecahedron,* arises from dualizing the black cuboctahedron. Since the corners of the cuboctahedron are all alike, the faces on the rhombic dodecahedron are all alike. Since four faces meet at each corner of the cuboctahedron, each face of the rhombic dodecahedron is quadrilateral. Since the faces on the cuboctahedron are of two sorts, there are two sorts of corners on its dual: three faces meet at each corner arising from a triangle, and four faces meet at each corner arising from a square. In an entirely similar way the principle of duality pairs the black icosidodecahedron with the white *rhombic triacontahedron.* You have already seen these rhombic solids diagramed on page 9.

The dual of the truncated tetrahedron is a triangular dodecahedron called the *triakis tetrahedron* because it looks like a tetrahedron with three-sided pyramids on its faces. Notice that again each Archimedean edge dualizes to a perpendicular edge that bisects it. But the new edge is not necessarily bisected by its parent.

A procedure for finding the shape of the face on any Archimedean dual is shown below, for the case of the truncated tetrahedron. First, find the shape of the polygon that forms the cross-section obtained by truncating a corner of the Archimedean as far as the midpoints of the edges that meet there. Then inscribe that polygon in a circle and draw tangents to the circle at the vertices of the polygon. This procedure wins validity from the fact that any Archimedean solid can be inscribed in a sphere. Notice another consequence of that fact: any Archimedean *dual* can be *circumscribed* to a sphere, which will touch the midpoints of all its faces.

Apply these ideas now to the snub cube and the snub dodecahedron. Since five faces appear at each identical corner, their duals are faced with identical pentagons. The twenty-four corners of the snub cube father the twenty-four-faced solid called the *pentagonal icositetrahedron,* and the sixty corners of the snub dodecahedron father the *pentagonal hexecontahedron.*

Looking back at the Archimedean solids (see p. 46), you recall that each has the same symmetry as a tetrahedron, a cube, or a regular dodecahedron, except the snub solids, which have no reflection planes. The duals to these solids all have the symmetries of their parents, as you have just noticed for the rhombic solids and

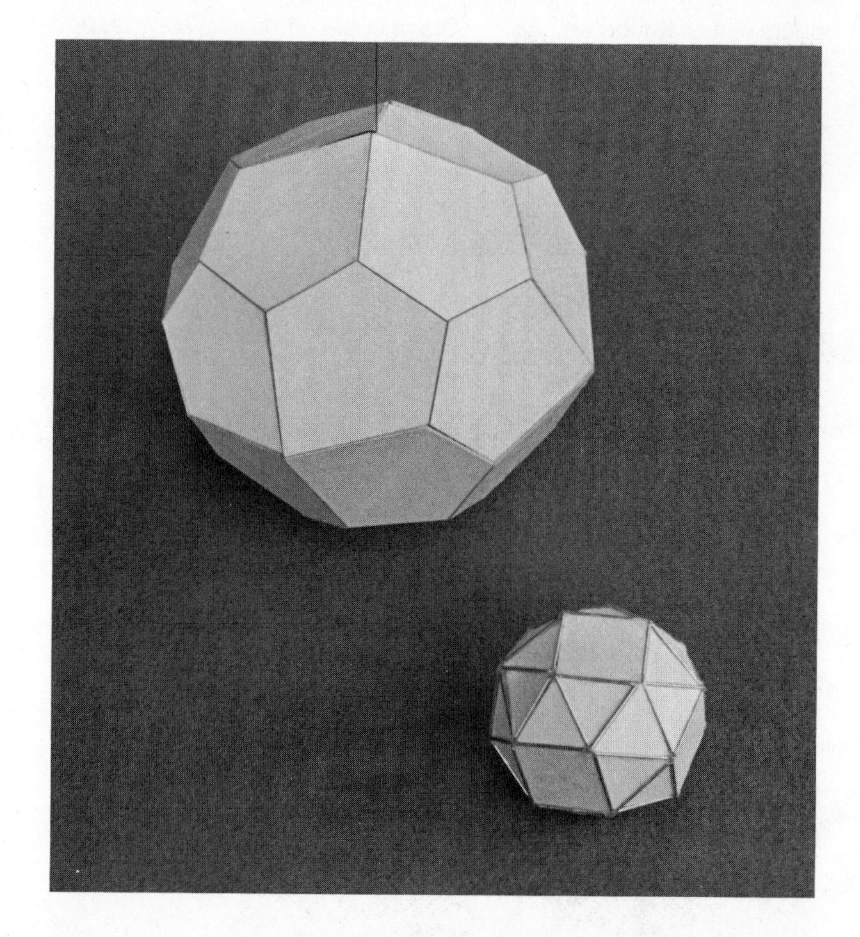

the triakis tetrahedron. Correspondingly, the symmetries of these pentagonal solids lack planes of reflection, and therefore they occur in left-handed and right-handed forms.

These solids approximate so closely to a spherical appearance that it is worthwhile recalling that they approximate to spheres in two different ways. In the Archimedean solids the corners fall on the surface of a sphere that is outside them; in their duals the centers of the faces fall on a sphere inside them. Since the faces on an Archimedean dual all have the same size and shape and stand at the same distance from the center of the solid, the solid is as likely to come to rest on any one of its faces when it rolls on a flat surface. Like the Platonic solids, therefore, the Archimedean duals are all suitable for dice, as the Archimedeans themselves are not.

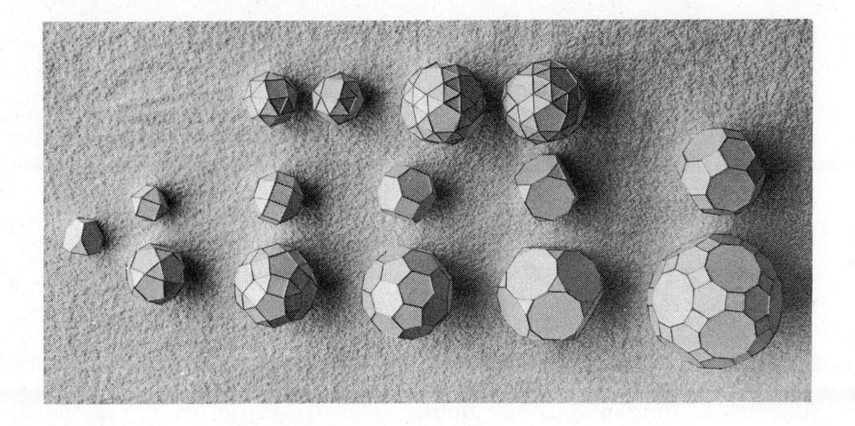

In order to pair off their duals with the thirteen Archimedean solids shown above, remember that: (1) a *face* on the dual will have the same number of sides as the number of faces that meet at a *corner* of its parent; (2) at a *corner* on the dual the same number of edges will meet as the number of sides on a *face* of its parent. The conventional names of these duals appear below.

| *Archimedean* | *dual* |
|---|---|
| truncated tetrahedron | triakis tetrahedron |
| truncated cube | triakis octahedron |
| truncated octahedron | tetrakis hexahedron |
| cuboctahedron | rhombic dodecahedron |
| small rhombicuboctahedron | trapezoidal icositetrahedron |
| great rhombicuboctahedron | hexakis octahedron |
| snub cube | pentagonal icositetrahedron |
| truncated dodecahedron | triakis icosahedron |
| truncated icosahedron | pentakis dodecahedron |
| icosidodecahedron | rhombic triacontahedron |
| small rhombicosidodecahedron | trapezoidal hexecontahedron |
| great rhombicosidodecahedron | hexakis icosahedron |
| snub dodecahedron | pentagonal hexecontahedron |

When you trim more and more off the edges of the white cube until the cubic faces disappear, the black rhombic dodecahedron arises. Such trimming could be called "truncating the edges." As you might expect, truncating the edges of an octahedron also will produce a rhombic dodecahedron. Similarly, truncating the edges of the white regular dodecahedron, or an icosahedron, will produce the black rhombic triacontahedron. The hexagons describing the black faces produced in the course of these truncations are never regular hexagons, and therefore the intermediate solids are not semiregular.

The truncations of corners and edges described so far all have one property in common: they attack in the same way all similar features of the solid. Consequently, each truncated solid has the same symmetry as its parent: no element of the original symmetry

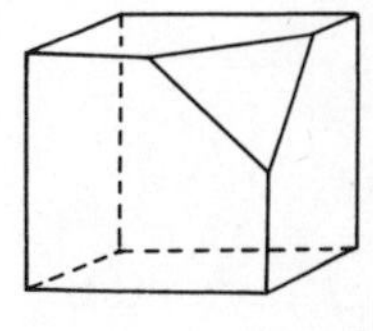

is destroyed. You can engage in a useful exercise, which will cultivate your abilities in visualizing spacial relations and in specifying symmetries, by truncating some but not all of the similar features of a solid and asking yourself what symmetry remains and what has been destroyed.

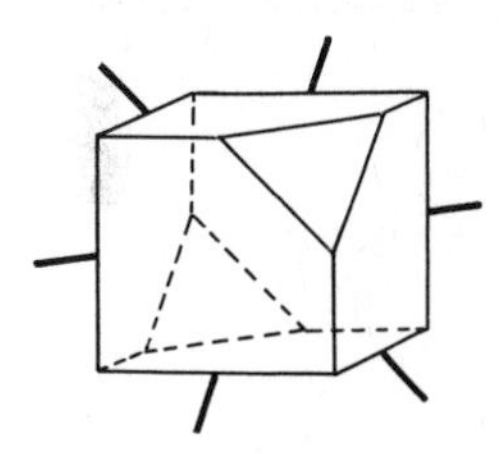

Truncating only one corner of a cube, for example, leaves one of its threefold axes of rotation unmolested but destroys the other three threefold axes. All fourfold and all twofold axes disappear. Only three of the nine planes of reflection remain. Truncating the opposite corner as well restores the three twofold axes that are perpendicular to the surviving threefold axis.

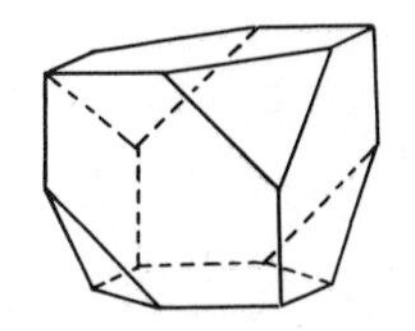

When four nonadjacent corners of a cube are truncated, the remaining symmetry is that of a regular tetrahedron, and it is interesting to see just how the truncation has degraded the cubic into the tetrahedral symmetry. The cube's three fourfold axes have degenerated into the tetrahedron's three twofold axes and the cube's own twofold axes have disappeared. The four threefold axes are still untouched. The six reflection planes through opposite edges of the cube remain, but the three planes parallel to the cube's faces have been destroyed.

Truncating the edges of a cube affords somewhat more difficult exercises of the same sort. Each of the solids shown above and outlined on the following page can be had by trimming some of the edges of a cube in the way shown at the top for one edge. Can you visualize which edges have been truncated to produce each solid? The eight solids are symbolized in the accompanying drawings by outlined cubes with heavy lines along the trimmed edges.

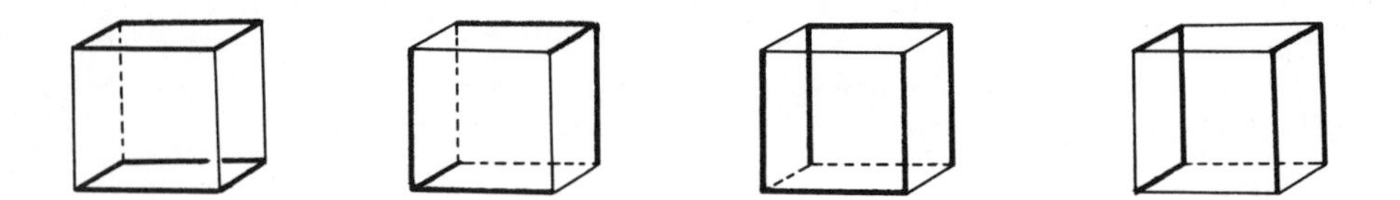

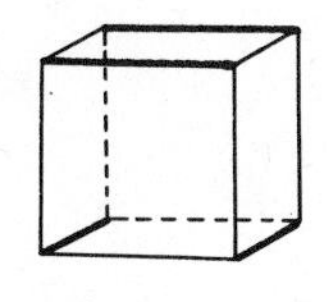

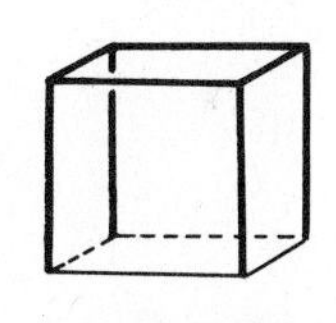

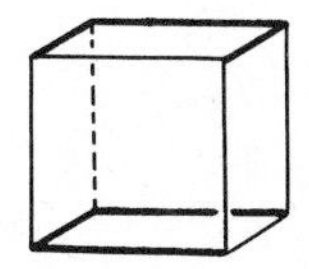

When you have paired off the pictured solids with the symbolic drawings, you can go on to specify the symmetry of each solid. You will probably find that your diagnoses can be made more easily from the symbolizing drawings than from the pictures of the solids. But if you were to build the solids, you would find that handling them is the best way to understand them. Notice that in two of these solids, which look much alike, all planes of reflection have been destroyed: they are mirror images of each other.

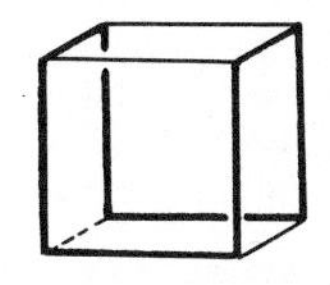

Around the beginning of the seventeenth century Johann Kepler, whose geometrical curiosity ran parallel with his astronomical interest, pointed out a class of solids having much in common with the Archimedeans. Any prism whose sides are squares, capped at top and bottom by a regular polygon, is semiregular: its faces are regular polygons of two sorts, and its corners are all alike. The same can be said of any antiprism whose sides are equilateral triangles. Thus Kepler enlarged, for the first time since the studies of the classical Greek geometers, the range of "mathematical solids" by adding the two series of **semiregular prisms** and **semiregular antiprisms.** Since the number of regular polygons is endless, both series are unending. Notice that the square prism is the familiar cube and the triangular antiprism is the familiar octahedron.

Like the Archimedean solids, the semiregular prisms and antiprisms can be inscribed in spheres. Hence the shapes of the faces on their duals can be found by the same procedure as that used for the Archimedean duals (see p. 51). The solids dual to the prisms are triangular *dipyramids.*

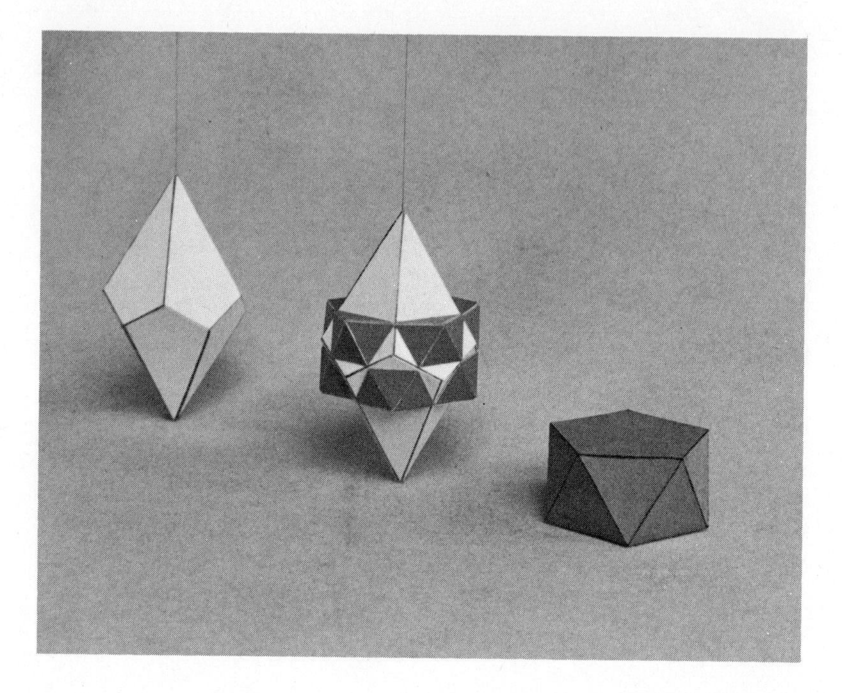

Since four faces meet at each corner of an antiprism, the faces on its dual are quadrilaterals. The faces have a kitelike shape: the solids are called *trapezohedra.* Like the Archimedean duals, and for the same reason, the duals to the prisms and antiprisms can all have spheres inscribed in them that will touch the centers of all their faces.

This fact suggests a recipe for making fair dice with any number of digits on their faces. Evidently the number of faces on a dual is twice the number of vertices on the polygon that caps the prism or antiprism. Hence by repeating each digit on two faces, a die can be made to have the same number of digits as that polygon. For a final touch of elegance in the recipe, you can seek to put the repeated digits on opposite faces. Notice that the corners of the prisms that are made of even-numbered polygons and the corners of the antiprisms made of odd-numbered polygons come in opposite pairs. Hence the dipyramids will be suitable for dice with an even number of digits, and the trapezohedra for dice with an odd number.

With this model of a rhombic dodecahedron, laminated upon a cube with sheets of glass, take your leave of the convex mathematical solids and turn to an even more exciting geometrical discovery by Johann Kepler.

If you draw a regular pentagon and then draw lines between its vertices, skipping one vertex for each line, five lines carry you back to your starting point and make a closed figure. The figure is a new sort of regular pentagon: its five sides all have the same length, and the angles at its five vertices are all the same. But the sides cross one another, and you must keep clearly in mind that a "side" of the new figure means the full course of a line from vertex to vertex, not a part of a line from a vertex to a crossing or between crossings. The figure is called a *regular star-pentagon,* or sometimes a *pentagram.* It is very ancient; the Pythagoreans in the fifth century B.C. used it as a symbol of their brotherhood.

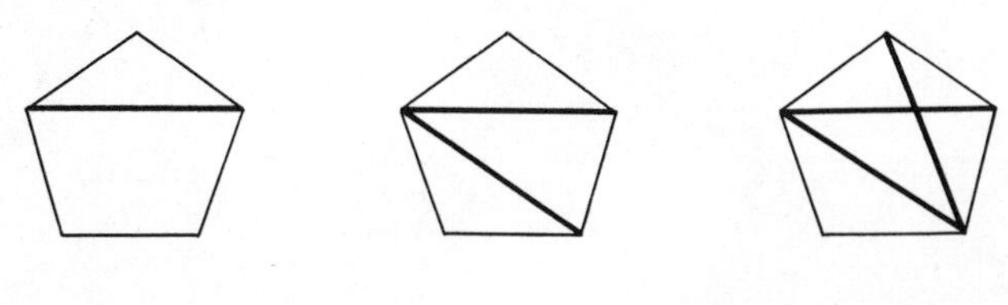

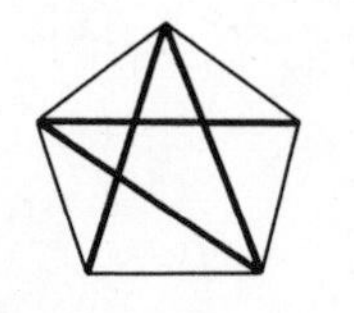

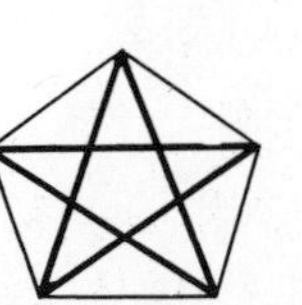

But it remained for Kepler to notice that twelve pentagrams can join in pairs along their sides and meet in fives at their vertices to form a closed solid. It fills all the requirements for a *regular solid:* its faces are all made of regular polygons of a single sort, and all its corners are alike. It differs from the Platonic solids only in the fact that it is not everywhere convex. It has twelve faces, twelve corners, and thirty edges. Each of the star-pentagonal faces has a five-sided pyramid projecting from its center. The solid has the full symmetry of the Platonic dodecahedron.

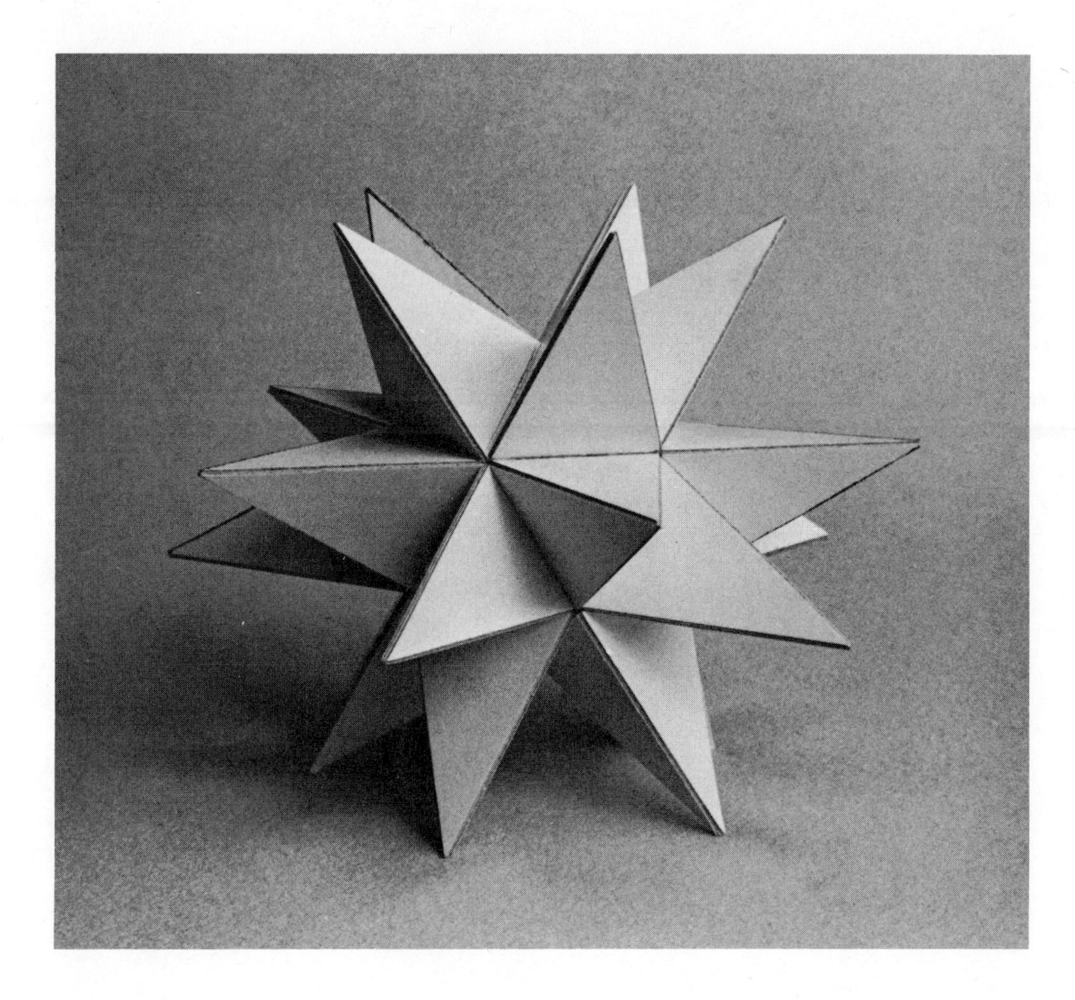

Kepler also noticed that regular star-pentagons can join in another way. When they meet at their vertices in threes instead of fives, they enclose a different solid that is also regular. Again the solid has twelve faces and thirty edges, but it differs from its predecessor in having twenty corners. Each of the star-pentagonal faces is somewhat obscured by a projecting rosette of five trigonal pyramids. Kepler's two new regular solids are now called the *small stellated dodecahedron* and the *great stellated dodecahedron* for reasons that will appear later.

Kepler's regular solids point to the rewards that can come from discarding the ancient Greek concern for convexity. Nevertheless, two centuries elapsed before Louis Poinsot discovered two more regular solids. Astonishingly, one of them is simply a dodecahedron, now called the *great dodecahedron,* whose faces are twelve ordinary pentagons. But here the pentagons intersect one another to produce a solid in which each pentagon carries an embossed star on its surface. The twelve pentagons meet along thirty edges at twelve corners. As on Kepler's solids, all the exposed parts of the faces, the parts needed to build a model of the solid, have the same shape.

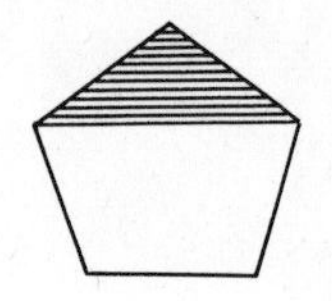

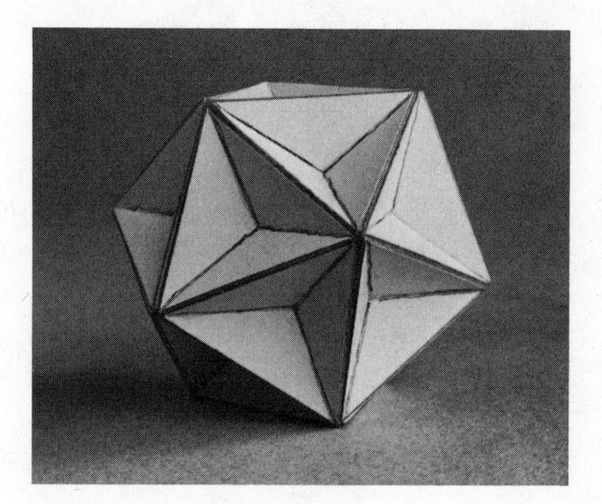

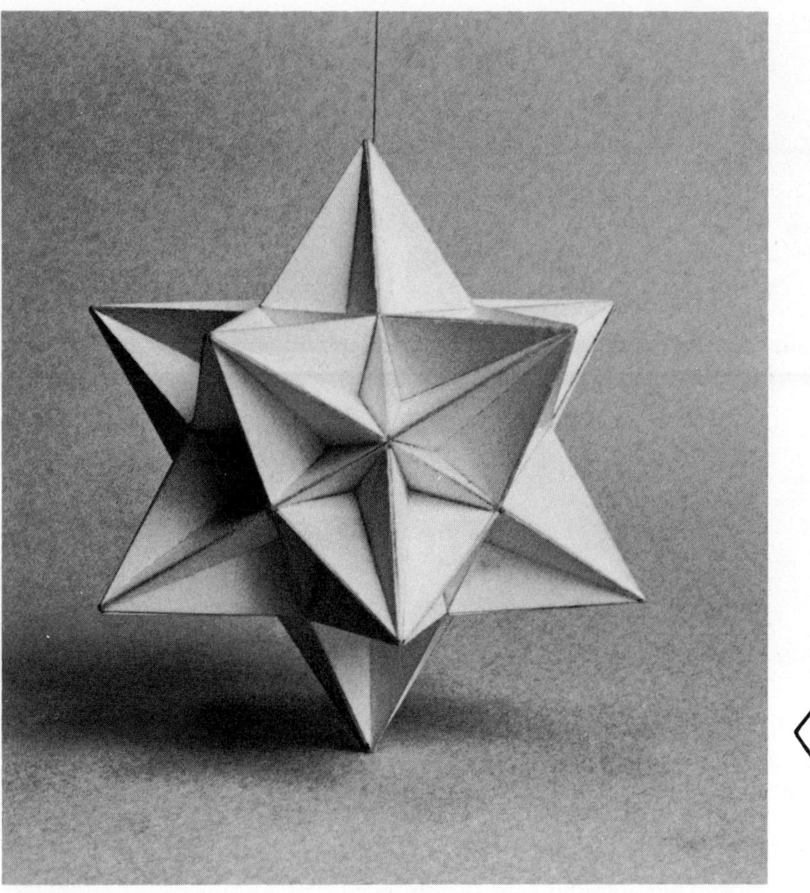

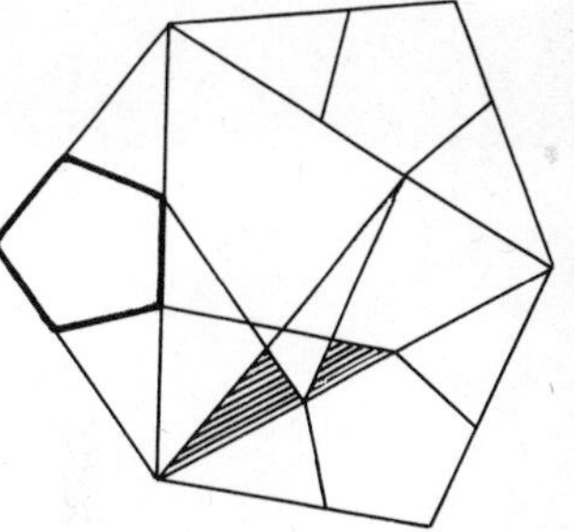

The second of Poinsot's regular solids is again astonishing: it is made of twenty intersecting equilateral triangles and is therefore called the *great icosahedron.* The triangles meet along thirty edges at twelve corners, as do the triangles in the familiar icosahedron. It is remarkable that twenty equilateral triangles can form regular solids in two such different ways. Building a model of the solid requires pieces with two shapes constructed as the diagram shows.

Collectively the four new regular solids are often called the **Kepler-Poinsot solids.** Will still more regular solids be discovered? It has been proved that no more are possible. The Platonic solids and the Kepler-Poinsot solids are exhaustive: if a mathematician refers to "the nine regular solids," he means these. The picture facing page 1 shows all of them together.

Look now at how the principle of duality applies to the Kepler-Poinsot solids. Remember that the Platonic solids were dual to one another. Interchanging corners and faces on any Platonic solid produced another Platonic solid. You could say that under the dualizing operation the Platonic solids form a closed set. And it turns out that the new regular solids also form a closed set under that operation: the Kepler solids are dual to the Poinsot solids. Notice one conspicuous aspect of their duality. The Kepler solids have convex corners and puckered faces, whereas the Poinsot solids have convex faces and puckered corners. On a model in which the great icosahedron interpenetrates its dual, a great stellated dodecahedron, the star-corners of the icosahedron lie above the centers of the star-faces of the dodecahedron.

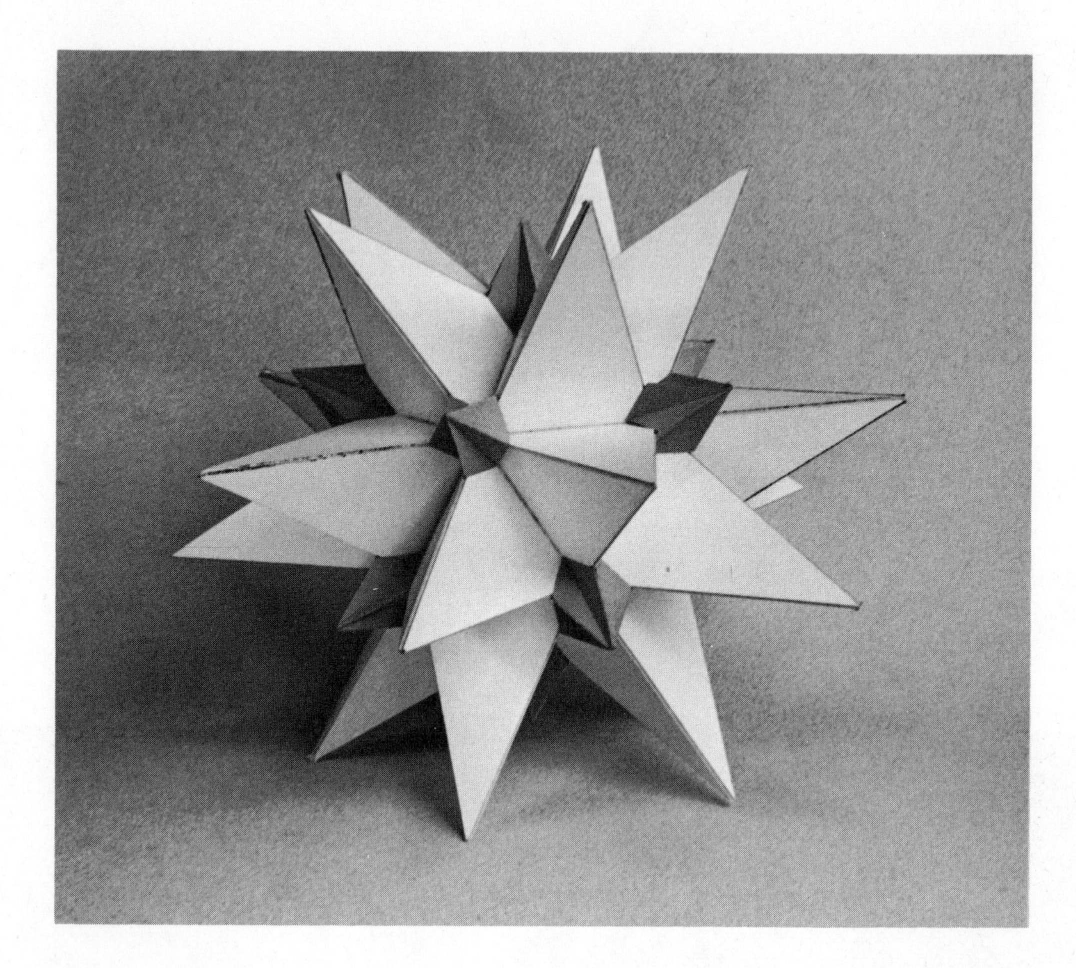

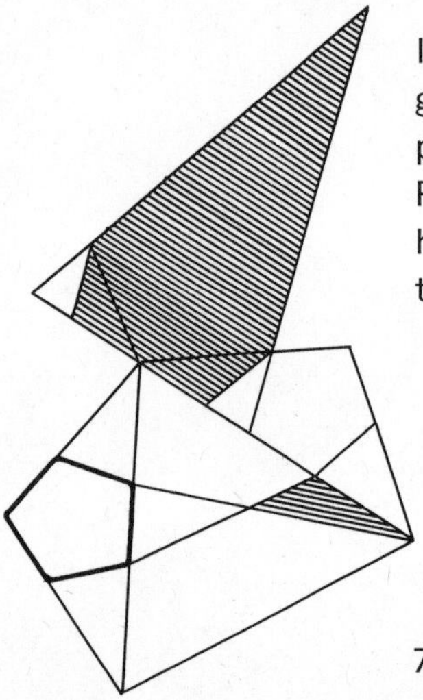

In a model showing the duality of the great icosahedron and the great stellated dodecahedron, the dual edges of the two solids are perpendicular bisectors of each other, like the edges of the paired Platonic duals (see p. 9). But here the intersections of the edges are hidden inside the model. The diagram shows how to construct the two sorts of exposed surfaces.

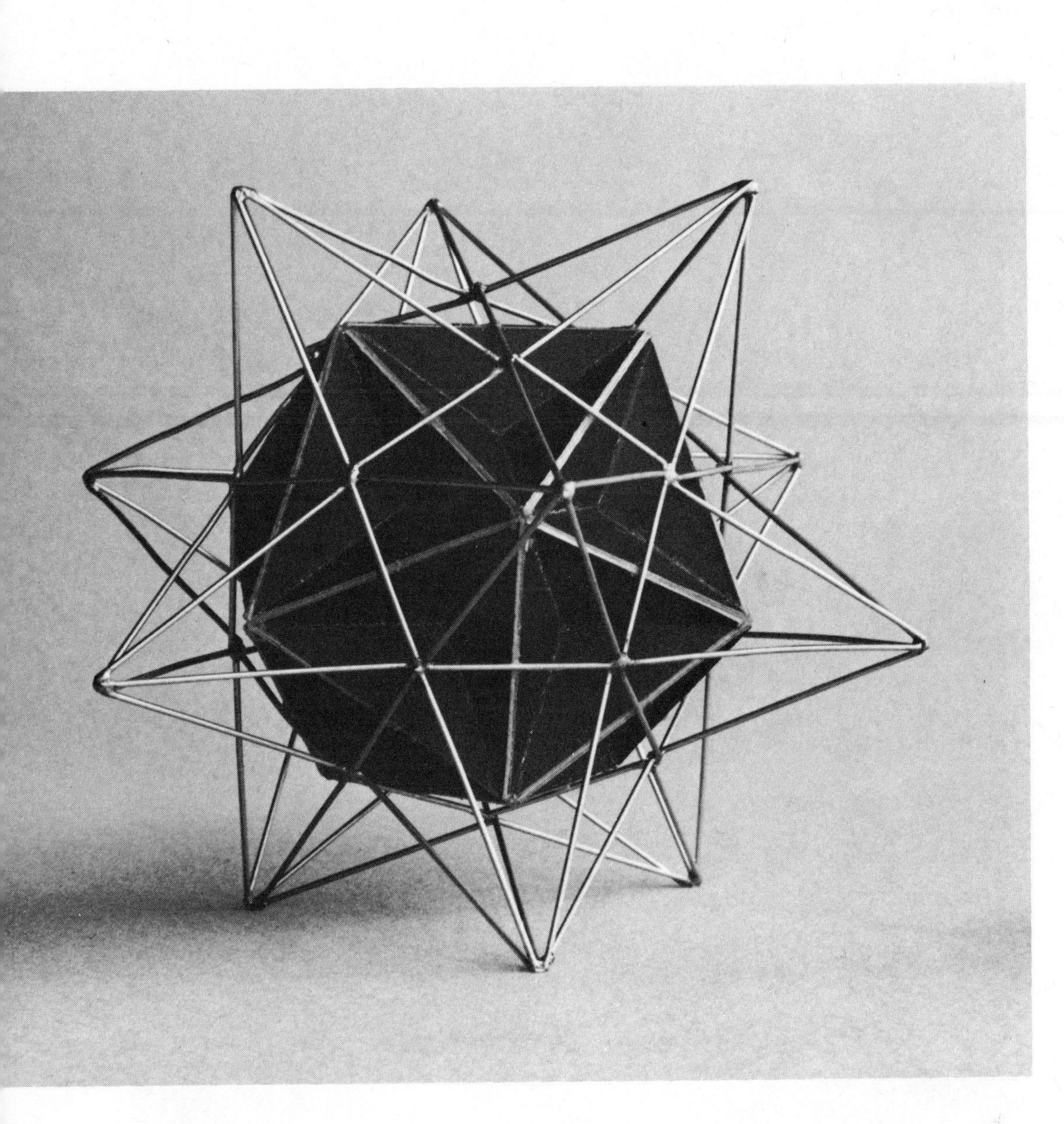

The great dodecahedron lies entirely inside its dual, the small stellated dodecahedron, when their sizes are chosen so that their edges intersect. Again the dual edges are perpendicular bisectors of each other.

The method used to produce the regular star-pentagon can be applied to other regular polygons with interesting results. The regular star-hexagon is well known as the Star of David. Unlike the star-pentagon, the Star of David can be decomposed into two regular polygons: it consists of two interpenetrating triangles. Proceeding to the regular seven-sided polygon, you can draw two regular star-heptagons by skipping one vertex for each new side and by skipping two vertices. Neither of these star-polygons is a combination of simpler polygons. Star-octagons again come in two sorts, obtained by skipping one or two vertices for each side. One of the polygons decomposes into two squares, but the other does not decompose.

Like the regular polygons, the regular star-polygons can form the caps of semiregular prisms and antiprisms. Since the sides of the star-polygons intersect one another, so do the squares and triangles on the prisms and antiprisms capped by stars. On the antiprisms the sides of the component triangles intersect one another not only where they meet the sides of the stars but also elsewhere. Clearly the star-polygonal prisms and antiprisms form again two infinite sets of semiregular solids. Indeed, proceeding to star-polygons with more vertices, you find these solids proliferating faster than the simple prisms and antiprisms because there are more and more

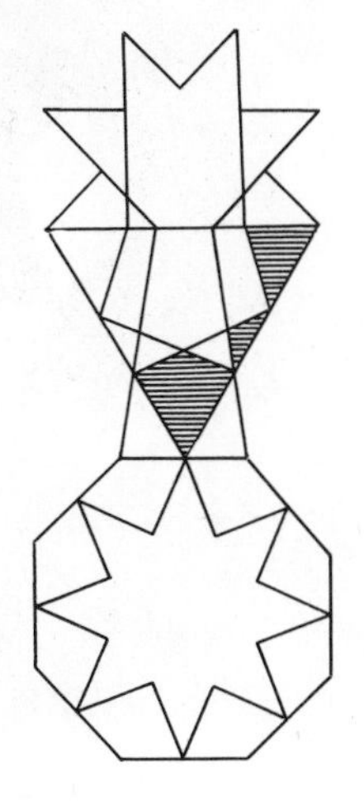

regular star-polygons with the same number of vertices, as the heptagonal and octagonal prisms left, and antiprisms right, remind you. Notice a curious difference between the antiprisms with star-caps and those with the more usual polygonal caps. When you look straight down on them, the corners of the usual antiprisms fall between one another, whereas the corners of the star-capped antiprisms, like those on the prisms, fall directly above one another. Hence the term "antiprism" here may seem a questionable use of language. With the antiprisms flowing from star-polygons drawn by skipping two vertices, the language becomes appropriate again.

As you might expect, the star-polygons that decompose into simpler polygons form parents for prisms and antiprisms that decompose into simpler ingredients. The prism fathered by the Star of David consists of two ordinary triangular prisms interpenetrating each other, and the corresponding antiprism decomposes into two interpenetrating regular octahedra. The decomposing star-octagon forms caps for two cubes, interpenetrating in a symmetrical arrangement that you may be interested to compare with the two other symmetrical compounds of two cubes shown on page 34. The corresponding antiprism decomposes into two square antiprisms.

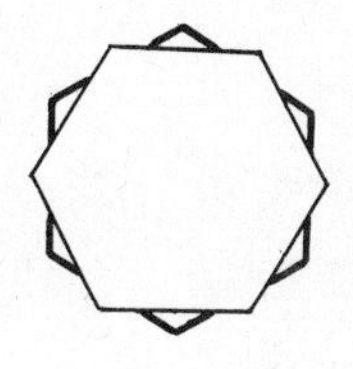

What can you expect as you travel further along the series of star-polygons? In general N distinguishable regular star-polygons can be made with either 2N + 3 or 2N + 4 sides. If the number of sides is a prime number such as seven, none of those star-polygons will decompose. If the number of sides is an even number such as eight, the star-polygon drawn by skipping one vertex for each side will surely decompose into two ordinary regular polygons with half that

number of sides. Others also may decompose. But when the number of sides is greater than six, there will always be at least one regular star-polygon that does not decompose. Of the three star-decagons pictured above, one decomposes into two regular pentagons, another into two regular star-pentagons, but the third is a new non-decomposing polygon. Of the four star-dodecagons diagramed, one is a compound of two hexagons, another of three squares, a third of four triangles, and the last does not decompose. The nondecomposing star-octagon and star-decagon appear as faces on some solids soon to be seen.

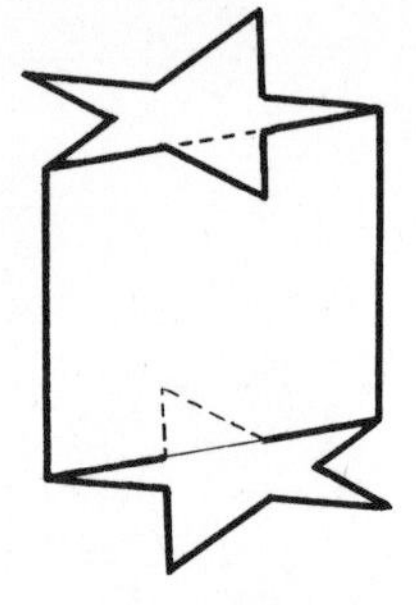

Conceivably you could build, upon star-polygons, semiregular "prisms" or "antiprisms" in which the squares or triangles pass through the solid from one side to the other as the diagrams suggest. What happens when you do this? Two opposite sides of each square meet sides of the two star-polygons, but the other two sides of the square meet nothing, and hence parts of the square faces enclose no inner space. It seems inappropriate to call such a figure a "solid." Playing the analogous game with equilateral triangles, however, you find that one side of each triangle meets a side of a star-polygon and the other two sides form edges with sides of two other triangles. The solid is truly a solid, whose faces everywhere enclose a volume. Here is yet another infinite series of semiregular solids.

Since each star-polygonal prism and antiprism is semiregular, and can be inscribed in a sphere touching all its corners, the shape of the face on its dual can be found by the method described on page 51. As on the duals to the more familiar prisms and antiprisms, the shape is triangular for the one and quadrilateral for the other. The star-polygons dualize to star-corners, as with the Kepler-Poinsot solids, and the edges of a dual perpendicularly bisect those of its parent.

Examine what happens when the faces of an octahedron are extended in their own planes. Each face meets the extensions of some faces that are not adjacent to it on the octahedron. By cutting one another off, the faces and their extensions construct new edges and new faces and so form a new solid. Each face becomes an equilateral triangle that is twice as big as the originating triangle on the octahedron. Kepler called the solid formed by the eight intersecting triangles the *stella octangula;* and today the operation of extending the faces of a solid until they intersect to form a new enclosing shape is called **stellating** the solid. In the guise of two interpenetrating tetrahedra, the stella octangula has already appeared on page 9.

Commonly, stellation is performed on solids all of whose faces are alike, as are the faces on the Platonic solids and on the Archimedean duals. But there is no reason to confine the operation to such solids. For example, when the truncated octahedron is stellated as shown above, a closed shape is obtained in which each hexagonal face has been expanded into a Star of David and each square face has become a square cross.

Early in the last century the celebrated French mathematician Augustin Cauchy showed that the Kepler-Poinsot solids can all be obtained by stellating either the regular dodecahedron or the icosahedron. The next few pages show how that comes about.

When the five faces surrounding any face on a regular dodecahedron are extended over that face, they enclose it under a five-sided pyramid. And when any one of the faces is extended similarly over all five of the faces that surround it, the face is expanded into a regular star-pentagon. Hence pyramids over all twelve of the faces, with triangular sides shaped like the points of star-pentagons, build a new solid that is a stellation of the regular dodecahedron and whose faces are regular star-pentagons. The solid is the *small stellated dodecahedron* (see p. 66).

Notice that the star-pentagonal faces of the small stellated dodecahedron can be extended further, by filling between the points of the star-pentagons to make regular pentagons out of them. The newly created pentagons meet to form new edges, burying the parental pyramids and producing the *great dodecahedron* (see p. 68). Finally, by extending still further each pentagonal face of the great dodecahedron into a star-pentagon, the dimples on that solid are buried under triangular pyramids, and the product is the *great stellated dodecahedron* (see p. 67).

Since each of the successive steps described on the preceding page is an extension of the faces on a single parental dodecahedron, all three of the solids obtained are stellations of the dodecahedron. In the picture above you see how the great dodecahedron can arise directly by extending the faces of a dodecahedron without passing through the intermediate stage of the small stellated dodecahedron.

dodecahedron
small stellated dodecahedron
great dodecahedron
great stellated dodecahedron

The dodecahedron can also be stellated directly into the great stellated dodecahedron. You will find it helpful to contrast the extension of *five* faces surrounding a *face,* producing a pyramid in the small stellation, with the extension of *three* faces surrounding a *corner,* producing the great stellation. Your examination will make clear that stellation cannot proceed further: you have reached an end.

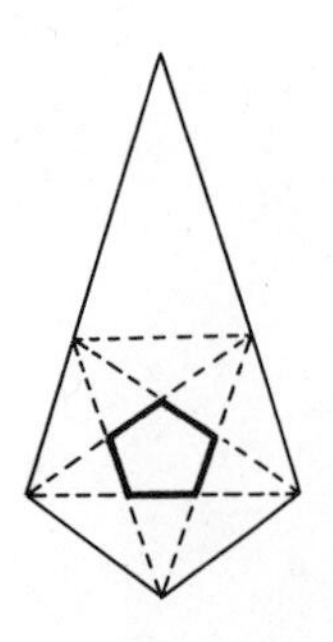

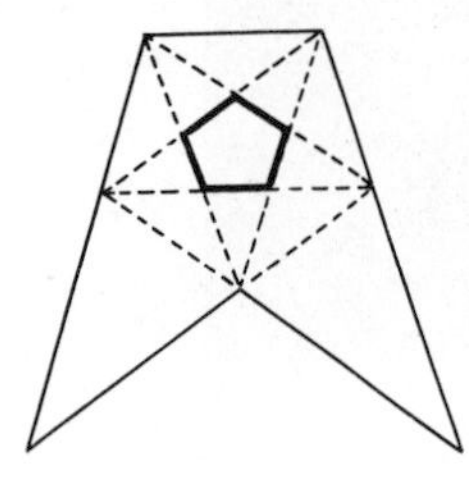

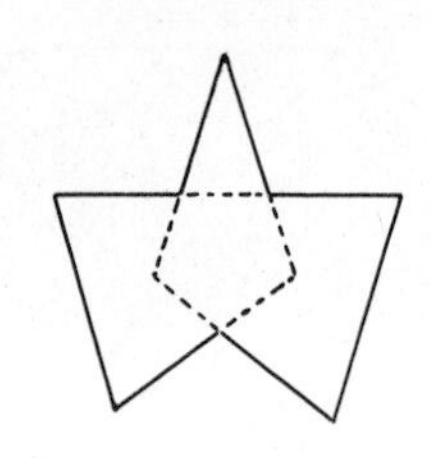

Some curious intermediate stellations of the dodecahedron arise from placing triangular pyramids over some but not all of the dimples on the great dodecahedron. Since the dimples are disposed as the faces on an icosahedron, four pyramids can extend each face of the dodecahedron into a blade and enclose a solid that comes in right- and left-handed forms. Eight pyramids produce a stellation that combines the other two and has three planes of symmetry. In another stellation with right- and left-handed forms, four of the dimples of the great dodecahedron appear.

The stellations of an icosahedron that treat each face alike are even more numerous than those of the dodecahedron: a systematic analysis has pointed out fifty-nine. Among them are compounds of five tetrahedra (see p. 39), ten tetrahedra, and five octahedra. Pictured here is the "first stellation," in which the three triangles surrounding each triangle are extended to cut one another off in a hexagonal shape that is not regular.

By continued extension, however, the faces finally cut one another off in equilateral triangles, as Cauchy discovered. Thus the fourth of the Kepler-Poinsot solids, the great icosahedron, turns out to be a stellation of the familiar icosahedron, just as the other three are stellations of the dodecahedron. Notice that stellation has expanded the original solid to a relatively enormous size. Each of the triangles forming the faces of the great icosahedron is cocked in the opposite direction from its parental triangle on the icosahedron. As on its parent, its triangles come in opposite pairs, with the members of each pair oppositely cocked. Its symmetry is therefore the same as the symmetry of its parent.

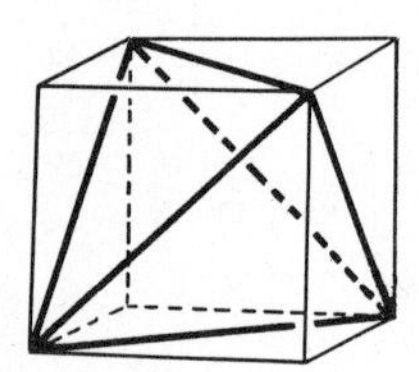

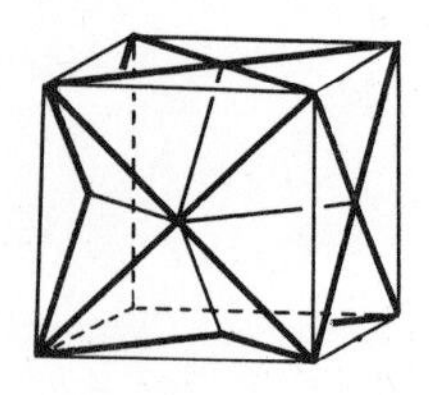

You may have noticed that the puckered pyramids of the great icosahedron project in the same directions as the convex pyramids on the small stellated dodecahedron. The relationship between the two solids can be established more precisely by observing that the corners and edges of the stellated dodecahedron form equilateral triangles that are disposed in the same way as the faces on the great icosahedron. The observation suggests another general way of constructing new solids: use a set of corners of a solid that lie in the same plane to form the vertices of a new polygon. Such polygons may outline new faces that join to enclose a new solid, even if the sides of the polygons do not fall along edges of the parental solid.

This operation on the parental solid is called **faceting** it. Faceting the small stellated dodecahedron has produced from it another regular solid whose corners and edges all correspond with those of its parent. In the much simpler case of the cube, faceting can produce the tetrahedron and the stella octangula, whose edges do not lie along the edges of the cube.

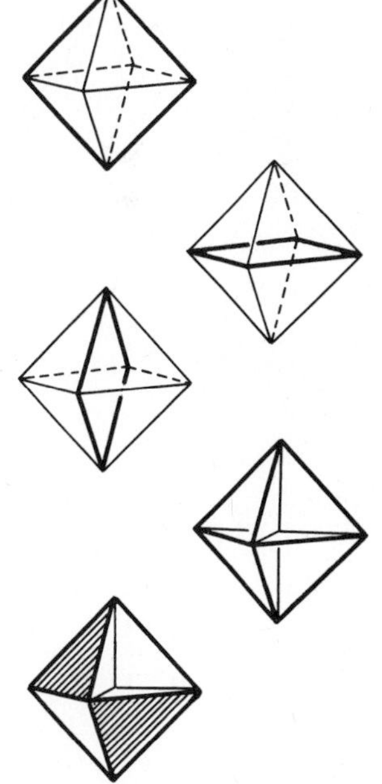

The most interesting new solids that arise from faceting are semiregular. They bear the same relation to the Archimedean solids that the Kepler-Poinsot solids bear to the Platonic solids. They are made of regular polygons of more than one sort, and all their corners are alike, but they are not convex. Thus they extend the Archimedean idea in much the same way that the star-polygonal prisms extend the idea of the more usual prisms.

The simplest example of these new solids comes from faceting the octahedron. Three sets of four coplanar corners can be chosen on the octahedron and taken as the corners of square faces, each cutting the octahedron in half. These three squares intersecting one another have already appeared on page 15. If you retain four faces of the octahedron, the three squares and four triangles form a solid on which a square and a triangle meet at each edge, which encloses a volume everywhere, and whose corners are all alike. You have already seen two semiregular heptahedra, the pentagonal and star-pentagonal prisms: here is a third.

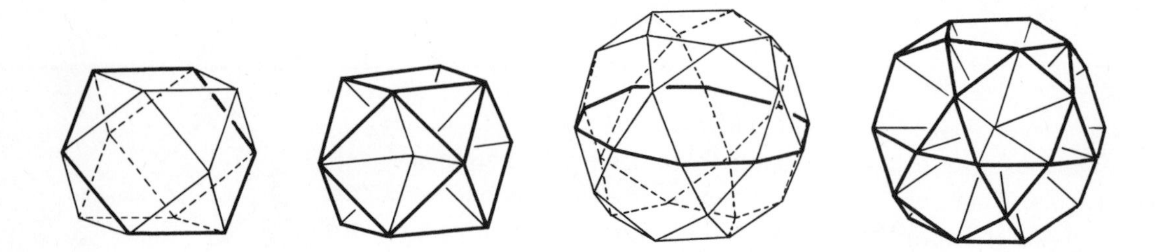

You can use the corners of a cuboctahedron in a similar way to define four intersecting regular hexagons, all passing through the center of the solid. Here you can retain either the original squares or the original triangles to enclose a semiregular solid. On one the faces comprise four hexagons and six squares and on the other four hexagons and eight triangles. Models are easily made when you notice that each side of a dimple is an equilateral triangle.

Analogous semiregular solids come from faceting the icosidodecahedron. Here the corners define six regular decagons, and either the original pentagons or the triangles will enclose a solid. Each side of a dimple is shaped like one point of a star-pentagon.

Turning to the small rhombicuboctahedron, you find its corners defining six regular octagons. Unlike the foregoing facetings, the octagonal faces do not pass through the center of the solid. A semi-regular solid can be produced by retaining either twelve of the original squares or six squares and eight triangles, in the latter case providing the first of these new solids whose faces comprise more than two species of regular polygons. It is worth your while to verify that two and only two faces meet at each edge, and that a circuit around any corner yields the sequence octagon-square-octagon-triangle. The analogous faceting of the small rhombicosidodecahedron produces twelve regular decagons, and the analogous semi-regular solids retain either thirty of the original squares or twelve pentagons and twenty triangles.

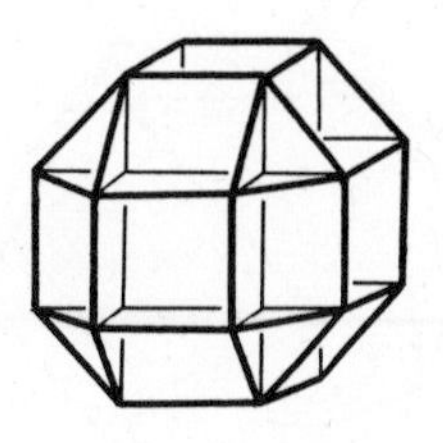

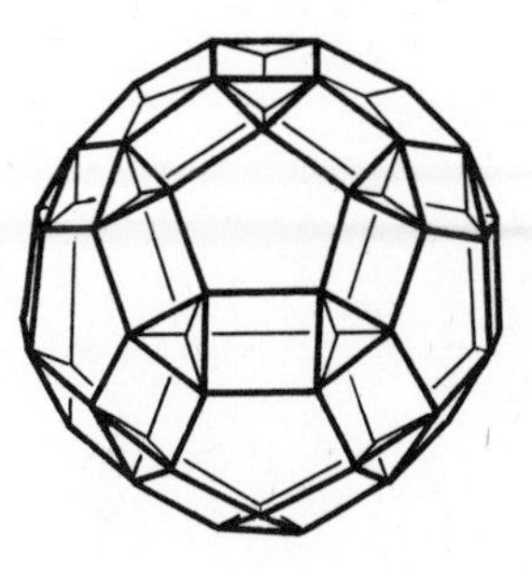

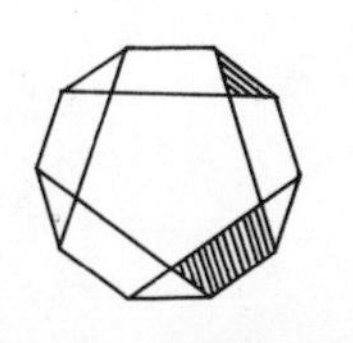

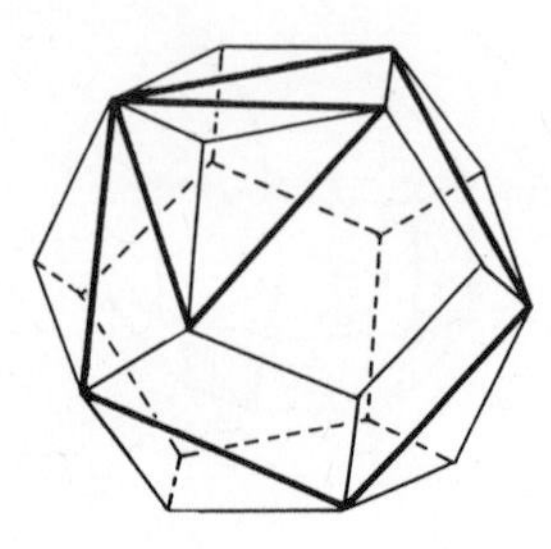

The dodecahedron can be faceted into twelve regular pentagons and twenty equilateral triangles that meet in pairs to form the edges of a very elegant semiregular solid. The disposition of its faces suggests that it be called the "great icosidodecahedron," but that name is commonly reserved for another solid, soon to be seen.

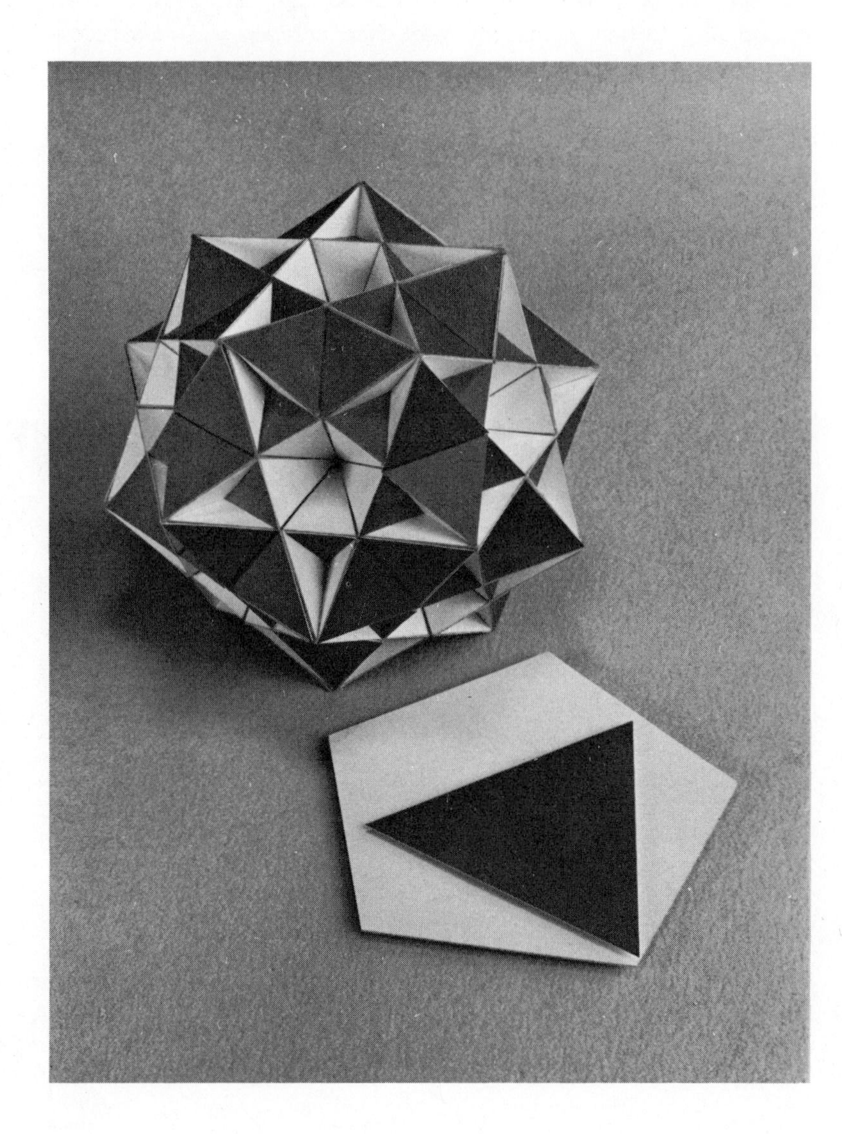

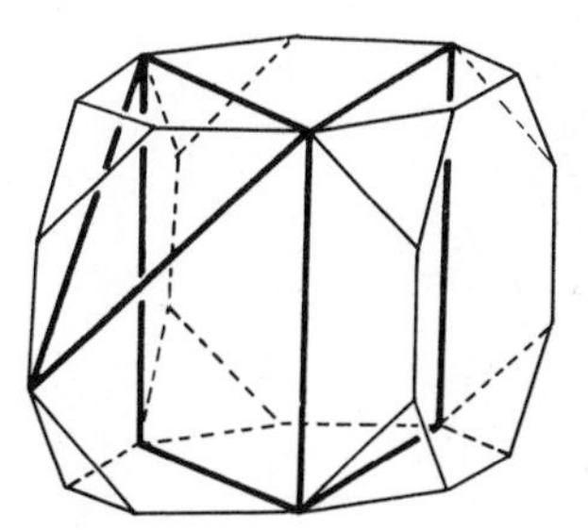

A semiregular solid, with eight triangular and eighteen square faces intersecting in a complicated way, arises from faceting the truncated cube. Its faces are disposed in a manner that suggests the name "great rhombicuboctahedron." On these two pages you see solids whose edges outline regular star-polygons, and you will not be surprised that star-polygons form the shapes of faces on many more semiregular solids.

If star-pentagonal faces were attached to the solid shown on page 98, three faces would meet along each edge, which is inadmissible in a semiregular solid. But on omitting the pentagons a valid semiregular solid arises in which a triangle and a star-pentagon join at each edge and three triangles and three star-pentagons meet at each

corner. It is easy to make a model of the solid when you notice that the exposed parts of the triangles are themselves little equilateral triangles. Alternatively, you can omit the triangles and retain the pentagons, obtaining a semiregular solid whose faces are twelve pentagons and twelve star-pentagons.

Truncating the Platonic solids produced many of the Archimedean solids. That fact suggests exploring what happens when you truncate the Kepler-Poinsot solids. As you start chopping off the corners of the great dodecahedron, little regular star-pentagons appear where the corners used to be. They preserve their regular shape as you truncate further, and when you have chopped far enough to convert the original pentagons to regular decagons you have produced a semiregular solid whose faces comprise twelve star-pentagons and twelve decagons. At each edge either two decagons or one decagon and one star-pentagon join, and at each corner one star-pentagon meets two decagons.

The truncation of the great dodecahedron can be continued until the star-pentagons meet at their vertices and the decagons are reconverted to smaller pentagons. The product is another semi-regular solid, with a pentagon and a star-pentagon joining along each edge and two pentagons of each species meeting at each corner. Because its faces comprise twelve pentagons of each species, it is called the *dodecadodecahedron.* This elegant solid can also be produced by faceting an icosidodecahedron with regular pentagons and covering with star-pentagons the open spaces that remain. A model is easily made by attaching star-pentagons to one another with three-faced dimples whose faces have the diagramed shape.

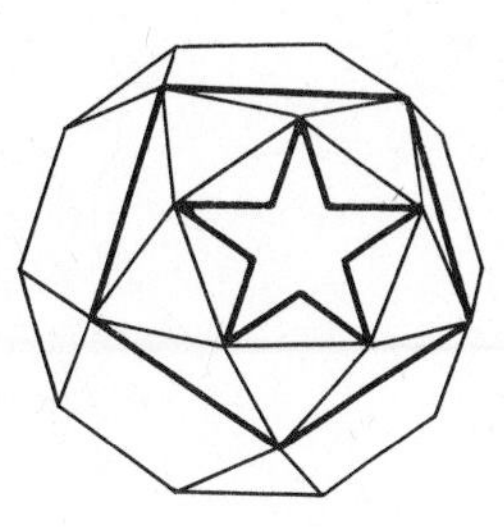

Turn now to the great icosahedron. Again truncating the corners replaces them with regular star-pentagons, and the operation can be successfully carried to the stage where the original triangles have been converted into regular hexagons. The sides of each hexagon join alternately with a side of a star-pentagon and a side of another hexagon. At each vertex a star-pentagon meets two hexagons. Unlike the truncation of the great dodecahedron, however, the truncation of the great icosahedron cannot be carried fruitfully any further: when the vertices of the star-pentagons meet, the hexagons have lost regularity. If you wish to make a model of this solid, a little thought about the method of construction of the great icosahedron shown on page 69 will avail.

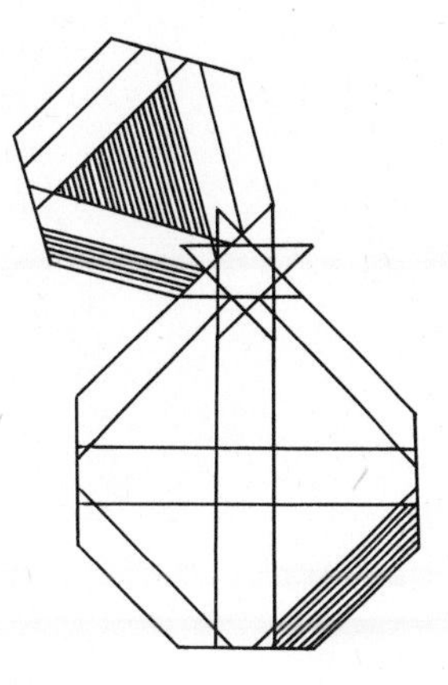

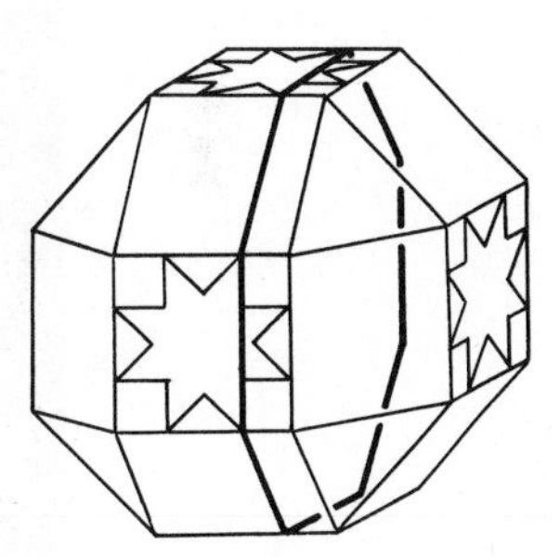

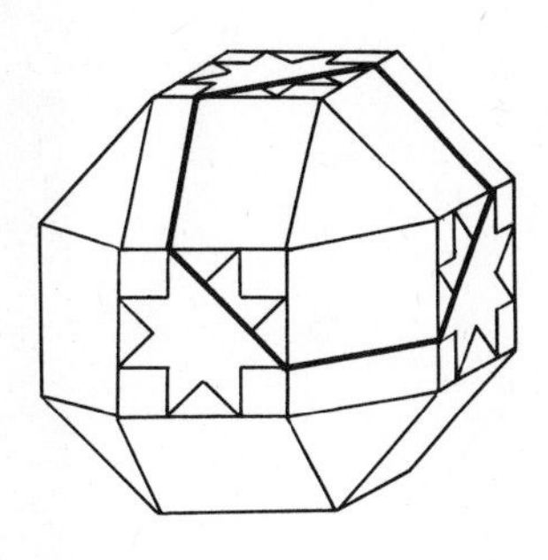

Inscribing regular star-octagons in six of the square faces of a small rhombicuboctahedron brings to notice a curious semiregular solid in which the star-octagons join regular octagons and regular hexagons.

Truncating the small stellated dodecahedron only leads back to the dodecahedron. Slicing it by planes perpendicular to its threefold axes of rotation, however, produces a solid in which twelve pentagonal faces intersect one another. The remaining faces could be regarded as "sectored triangles" and do not provide a semiregular solid. Truncating the corners at which those triangles meet replaces the corners with regular pentagons and converts the sectored triangles to sectored hexagons.

In a similar way, slicing a great stellated dodecahedron by planes perpendicular to its fivefold axes yields a solid whose faces comprise twelve pentagons and twelve sectored pentagons. When the corners where the sectored pentagons meet are cut off, little triangles appear, and the sectored pentagons become sectored decagons. But the sectored polygons do not qualify as regular polygons, and these solids cannot properly be called semiregular.

The edges of the solid shown on page 99 outline star-octagons. When star-octagonal faces are placed upon the solid, three faces meet along each edge. But when twelve of the eighteen square faces are removed, the solid becomes legitimately semiregular again, with each side of a star-octagon joining a side either of one square or of one triangle. To make a model of the solid, attach the star-octagons to one another with dimples made of three little squares, and then add the little equilateral triangles representing the exposed parts of the triangular faces.

Recalling that the star-octagonal solid was obtained by faceting the truncated cube, you can make an analogous star-decagonal solid by faceting the truncated dodecahedron. Each side of a star-decagon joins a side of either a pentagon or a triangle. Again a model is best made by attaching the star-decagons together by the dimples, which have the same shape as the dimples on the dodecadodecahedron (page 103), and then attaching little equilateral triangles.

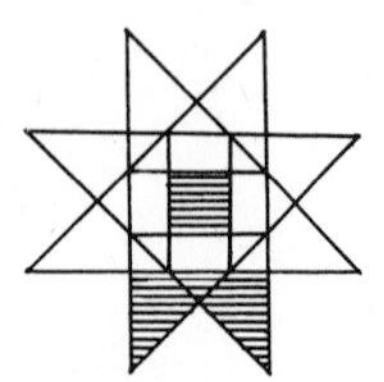

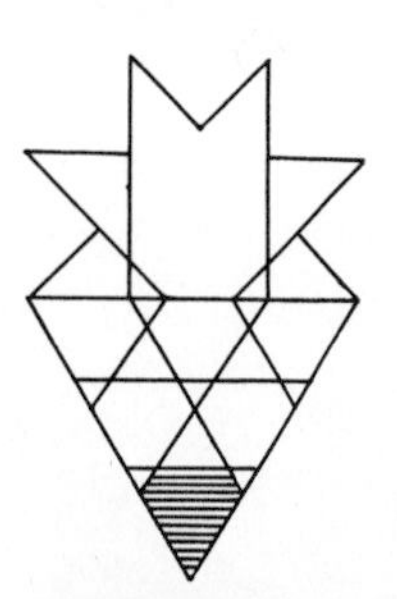

On all the semiregular solids with star-polygonal faces that have been shown so far, those faces have been fully exposed. Only the faces shaped as ordinary polygons have intersected one another along lines other than their sides. Turn now to some beautiful semiregular solids in which star-polygons also intersect one another.

The faceting of the small rhombicuboctahedron to yield regular octagons, already shown on page 97, makes a good starting point. If the vertices of those octagons are used to draw nondecomposing regular star-octagons, the star-octagonal edges can be taken as defining the edges of eight equilateral triangles. In the resulting semiregular solid, triangles and star-octagons alike intersect one another, and two star-octagons and a triangle meet at each corner. A model can best be constructed by attaching six crowns, made of the diagramed parts, to the faces of a cube.

An analogous semiregular solid, whose intersecting faces are shaped as twelve star-decagons and twelve pentagons, arises from faceting the small rhombicosidodecahedron. Here twelve crowns ornament the faces of a dodecahedron.

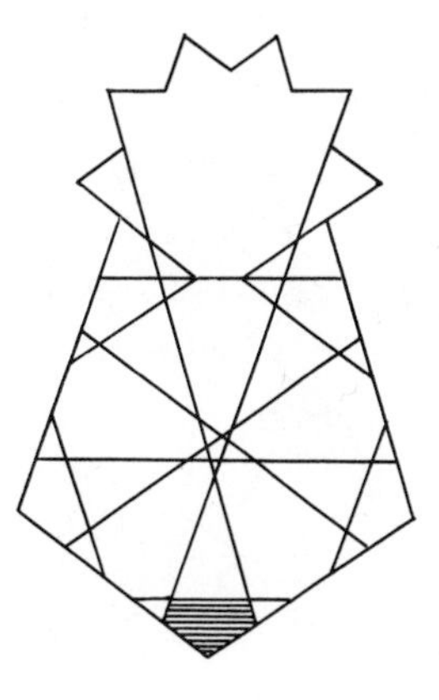

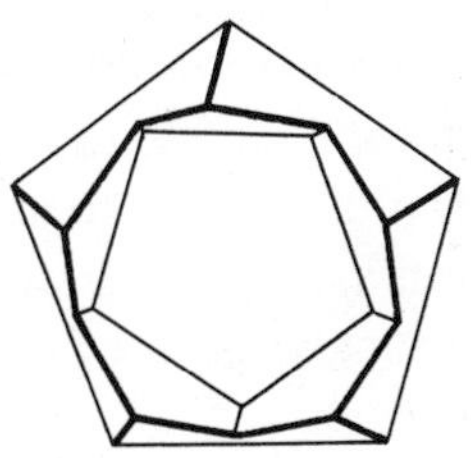

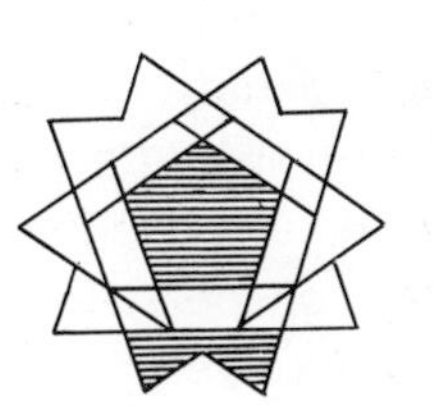

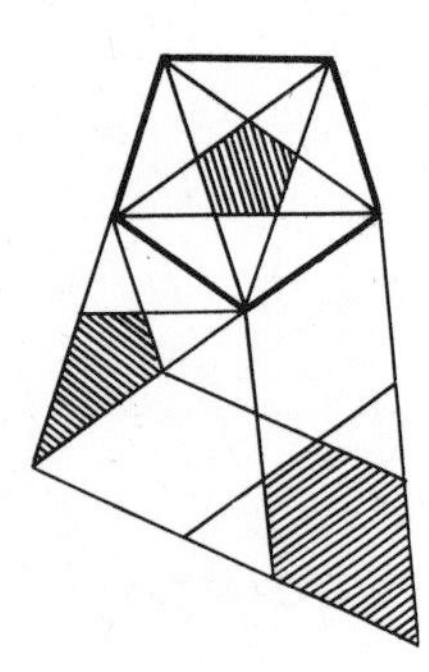

Of the nonconvex semiregular solids, surely one of the most beautiful is the *great icosidodecahedron.* Its twelve star-pentagons and twenty triangles all intersect; two of each species meet at each corner. The exposed parts of the triangles form five-sided dimples, and a star-pentagon cuts through each dimple to make an appearance as a little pentagon. The diagram shows how to find the three shapes needed to make the parts for a model. Notice that the corners and edges of the solid can be arranged on six nondecomposing star-decagons.

Six star-decagonal faces, of the sort outlined on the preceding page, could be made to pass through the center of the great icosidodecahedron. Choosing them for faces on a new solid, you can make that solid semiregular by discarding the twelve star-pentagons of the parental solid and retaining its twenty triangles. Then two star-decagons and two triangles meet at each corner. Three triangles cut through each of the large three-sided dimples between the star-decagons and form subsidiary dimples.

Alternatively, you can discard the twenty triangles and retain the twelve star-pentagons of the great icosidodecahedron. Then two star-decagons and two star-pentagons meet at each corner. It is worth remarking that this semiregular solid is made entirely of star-polygons, as Kepler's solids are made. Here each star-pentagon, cutting through a large five-sided dimple between the star-decagons, makes an appearance as a regular pentagon.

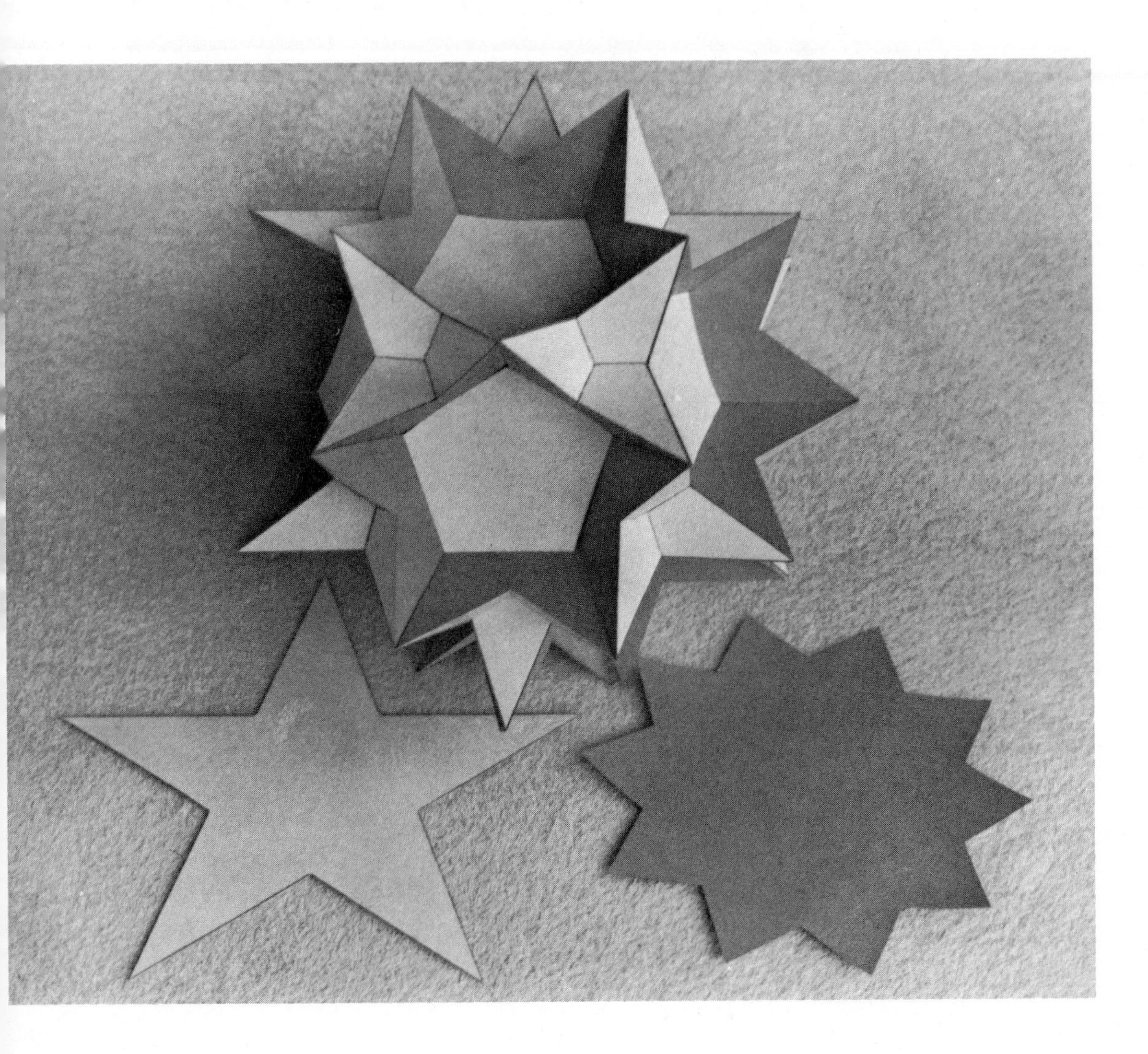

You have already seen how the principle of duality extends to regular solids and semiregular prisms when they are not convex. The nonconvex analogs of the Archimedean solids can be dualized in the same way. Again they are not dual to one another as the Kepler-Poinsot solids are. But again, since all their corners are identical, the faces on any dual have the same shape. And again, since each of these new semiregular solids can be inscribed in a sphere, the shape of the dual face can be found by the construction described on page 51.

Applying these principles to the solid shown above on the left, you obtain the dual solid shown at the right. The corners dual to the triangles are the convex corners of the three-sided pyramids; the corners dual to the star-pentagons are the star-corners at each of which five edges meet. The shape of the face is an irregular hexagon.

Recognize now that this solid is the "first stellation" of the icosahedron that has already appeared on page 92. This fact furnishes one example of an interesting principle that has more general application. The corners of the solid at the left all fall on its ten axes of threefold rotation symmetry. Hence the faces dual to those corners must be perpendicular to those axes. Since the solid has twenty

corners, its dual has twenty faces, and the convex solid with twenty faces perpendicular to threefold axes is the icosahedron. Thus the faces of the dual must be parallel to the faces of an icosahedron. Since they must all touch the surface of an inscribed sphere, they must form a stellation of the icosahedron.

Looking at the great icosidodecahedron and its dual, you see these principles at work again. Its triangles dualize to corners on three-sided pyramids; its star-pentagons dualize to star-corners projecting from the middle of five-sided dimples. Since the corners of the great icosidodecahedron fall on its fifteen twofold axes, the faces of its dual are parallel to the faces of the dual to the Archimedean icosidodecahedron. In other words, the dual is a stellation of the rhombic triacontahedron shown on page 50.

The shape of the face on the stellated rhombic triacontahedron dual to the great icosidodecahedron is a rhombus, different from that of the triacontahedron itself, with acute angles appearing at the convex corners and obtuse angles at the star-corners.

Another semiregular solid whose thirty corners lie on its axes of twofold rotation symmetry is the dodecadodecahedron (see p. 103). Again, therefore, its dual must be a stellation of the rhombic triacontahedron. Superficially the dual looks a little like Kepler's small

stellated dodecahedron; it consists of twelve five-sided pyramids.

When you think of this solid as dual to the dodecadodecahedron, however, something seems lacking. You recognize at once the tips of the convex pyramids as the corners dual to the regular pentagonal faces of its parent. But where are the expected star-corners that are dual to the parental faces shaped as star-pentagons? To answer this question, examine on the next two pages how the process of stellating the rhombic triacontahedron might proceed.

Starting with the rhombic triacontahedron at the left, and extending each face over its immediate neighbors, you bury all the faces under shallow four-sided pyramids. The new face on this first stellation has a double-winged shape. Extending each new face further, so as to fill in the two reentrant angles between the wings, makes the face into an irregular hexagon. But the side of the added filling does not meet any other side, and therefore the extension does not enclose a volume: the face must be prolonged further, as the wires show. The face then becomes a thin rhombus, again different from the rhombus on the triacontahedron itself, and its sides join with those of its fellows to form the edges of the conspicuous pyramids. The edges extend into the solid, meeting again in corners at the obtuse angles of the rhombic faces. Thus those obtuse angles form star-corners as in the stellation shown on page 118, but here the star-corners are located directly inward from the corners of the pyramids and cannot be seen without invading the model. You will not find this surprising when you remember that all the corners are dual either to pentagons or to star-pentagons lying parallel to one another in pairs. The dual corners must therefore lie in pairs on axes of fivefold rotation symmetry. The shape of one of the rhombic faces, and the invasion revealing its obtuse angles, appear below. Thus you are led to the remarkable idea that a solid may have corners that cannot be seen.

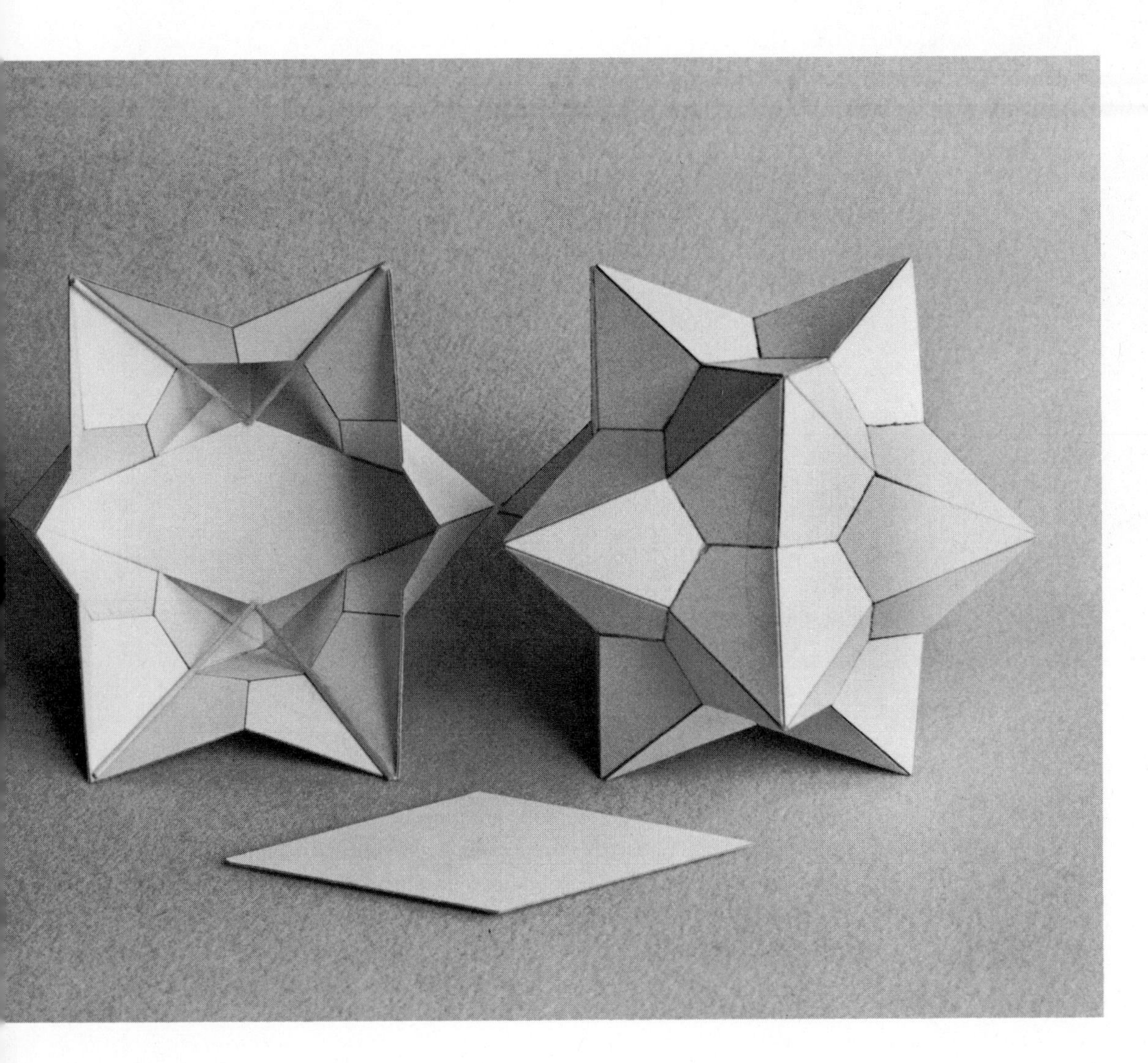

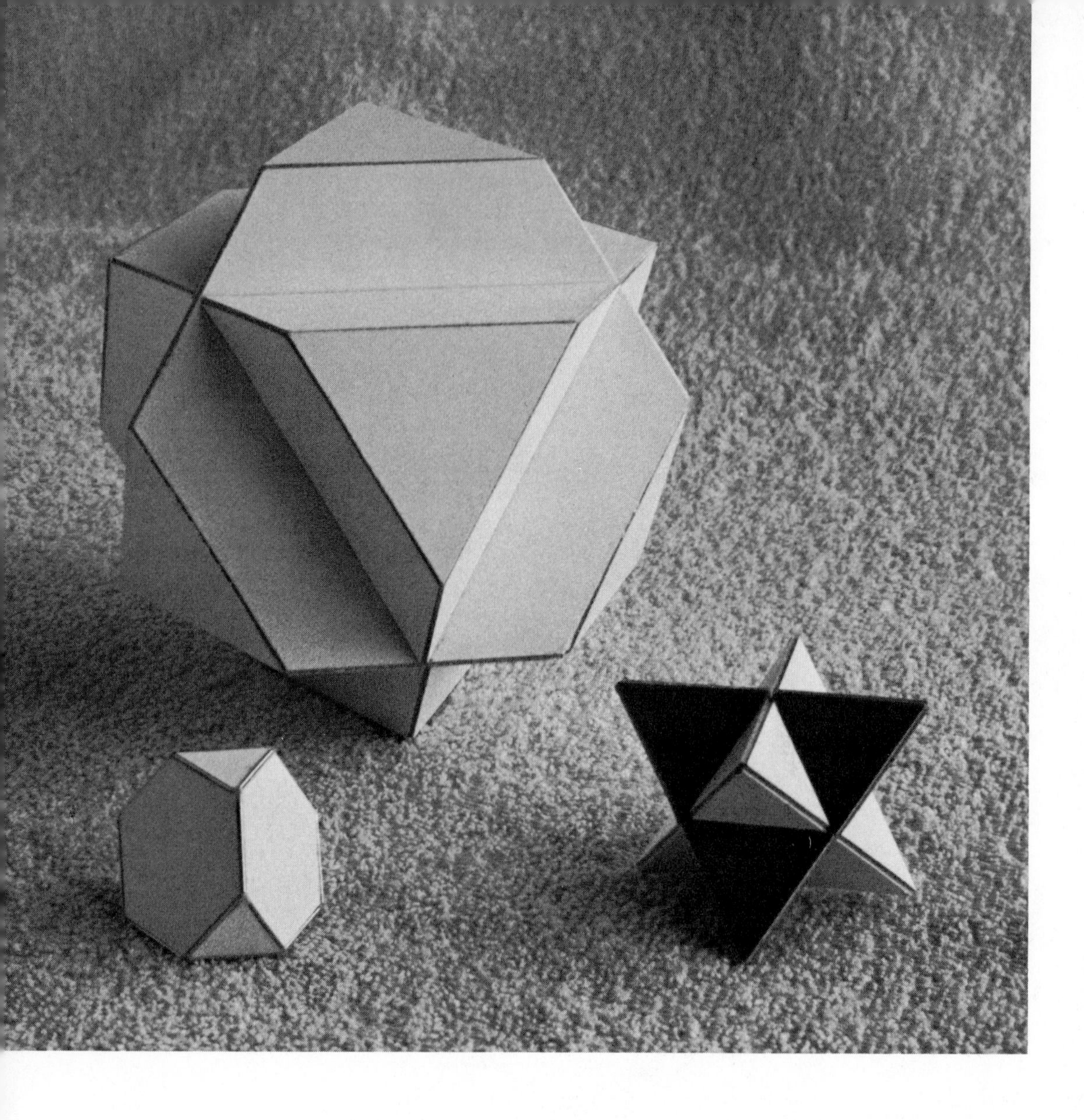

Are there still more solids that are semiregular? Notice that you have already seen nonconvex analogs of the Archimedean solids numbering more than the thirteen Archimedeans themselves. Assiduous search by mathematicians has brought to light fifty-three such analogs. It has been proved that the thirteen Archimedeans exhaust the convex possibilities, but there is so far no proof that the nonconvex possibilities have been exhausted by the fifty-three discovered.

If you are inspired to search for still more of these solids, you would do well to examine some possibilities that look attractive but do not qualify. Often you will be deceived by solids that turn out to decompose into simpler solids interpenetrating one another. The handsome solid on the preceding page has eight triangular and eight hexagonal faces; at each corner a triangle and two hexagons meet. It is a truncated stella octangula, and therefore it decomposes into two interpenetrating truncated tetrahedra.

For another example, recall your success in deriving the semiregular solid shown on page 108 by placing star-octagonal faces over the solid made of squares and triangles shown on page 99 and omitting twelve of its squares. You might be tempted to restore the squares and omit the triangles. But the resulting solid decomposes into three interpenetrating star-octagonal prisms.

Repeatedly in this book you have met and discarded figures whose faces do not meet others at all parts of their sides and so leave places where the faces enclose no volume. Now proceed in the other direction for a moment: examine some figures made of faces that enclose no volume anywhere. An assembly of faces forming a "polyhedron of volume zero" is sometimes distinguished from a solid by the name **nolid**.

Especially interesting are the nolids that might be called *regular nolids.* Each face has the shape of a single sort of regular polygon,

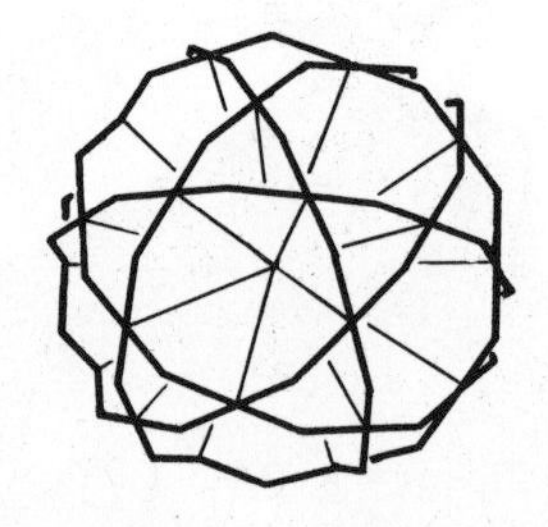

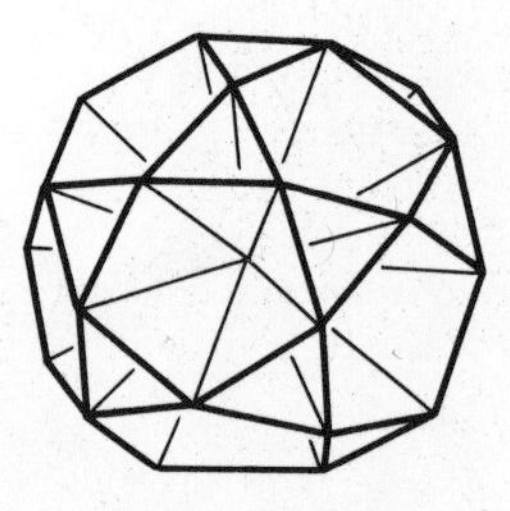

and all the corners are alike. When the faces immediately surrounding a corner are solely the one face appearing at the corner, the requirement that all corners be alike must be taken to mean that all corners are surrounded in the same way by other corners. Two such nolids on the opposite page are made of three squares, one of six squares, two of four hexagons, and one of three nondecomposing star-octagons. All have the full symmetry of a cube. Above are nolids made of triangles, hexagons, and decagons.

A curious nolid arises from using the twelve corners of a cuboctahedron as the vertices of four equilateral triangles. It has all the axes of rotation symmetry of a cube, but it lacks planes of reflection symmetry and therefore comes in left- and right-handed forms. No regular solid has this property.

The solids so far described can be placed in two classes by taking note of a conspicuous visual distinction. Most of them look the same from many different directions. But the prisms and antiprisms exhibit a unique "up-and-down" direction, and the compounds of two cubes on page 34 also have that character.

Now examine some more precise means for making the desired distinction. Each of the solids with the up-and-down character has one axis of rotation symmetry that is unique, not duplicated in another direction. This observation suggests examining what occurs when an axis of rotation symmetry is duplicated in some other direction.

If two twofold axes are chosen in two different directions, each of the axes requires that the other be repeated after a half-turn about it. Since each axis so produced makes the same requirement on all others, there must be twofold axes in all directions unless the directions of the generating pair are adroitly chosen. One such choice is to place the axes perpendicular to each other. In fact, if you prescribe three mutually perpendicular twofold axes, they will be self-consistent, entailing no more. But they are still not sufficient to define the desired distinction. A long narrow box has those axes of symmetry and still has a conspicuous up-and-down appearance.

Surprisingly, it turns out that a minimal prescription of the character in question can be made in terms of threefold rotation axes. When two such axes are arranged in space in the directions that they would take if they were threefold axes of a cube, rotation about either axis in two steps of 120 degrees carries the other axis through the positions of the other two threefold axes of a cube. Thus the four axes form a self-consistent set. And an object having at least that much symmetry cannot exhibit a conspicuous up-and-down direction.

Look next at what this minimal symmetry entails, by marking on one face of a cube some figure having no symmetry in itself, such as a comma. The cube serves merely as a convenient frame for the axes and a vehicle for the comma. Turning the cube about one of the threefold axes requires the comma to appear on two more faces. Turning it about the other threefold axes requires the final appearance of twelve commas. Looking now at that array of commas, you see that it has not only the four threefold axes used to construct it but also three twofold axes through opposite faces of the cube. You conclude that the four threefold axes are *necessarily* accompanied by three twofold axes, and *not necessarily* by any further symmetry. Anything having at least that much symmetry is called an **isometric object**. Since no reflection plane is required, the isometric wire figure below can be made right-handed or left-handed.

The wire figures on the preceding page lack planes of symmetry because their pyramidal components all twist about their threefold axes in one or the opposite sense. When the four pyramids are not twisted, the figure acquires six reflection planes. The resulting symmetry is that of the regular tetrahedron, already examined on page 19.

When the two wire figures of page 127 are combined into one exhibiting eight pyramids, four twisting in one sense and four in the other, the figure acquires three reflection planes, different from any of those on the preceding page. You have already seen the resulting symmetry in a stellation of the dodecahedron shown on page 91.

When the eight pyramids all twist in the same sense, the figure loses its planes of reflection again, but six new twofold axes spring into being. For the first time the figure exhibits the twofold axes of the cube itself.

Furthermore, its three former twofold axes now become the three fourfold axes that are especially characteristic of the cube. Thus the symmetry of the figure includes three fourfold, four threefold, and six twofold rotation axes, which compose all the rotational symmetry of the cube. But the figure still lacks reflection planes and can occur in either a left-handed or a right-handed form according to the sense in which the pyramids twist about their threefold axes.

This is the symmetry that you have found in the snub cube on page 45, and again in its dual on page 52. If the pyramids lose their twist and straighten into ordinary pyramids, the figure retains all its axes and acquires the six reflection planes shown on page 128 and also the three shown on page 129. In other words, it finally exhibits the full symmetry of the cube.

When the symmetry of the cube itself is included, there are five possible isometric symmetries that are subgroups of the full cubic symmetry. Arrangements of commas on a cube display all five: right-handed and left-handed arrangements appear when the symmetry lacks planes of reflection. The table may help you to correlate these ornamented cubes with the foregoing discussion.

| *symmetry* | *planes* | *2* | *3* | *4* | *page* |
|---|---|---|---|---|---|
| 1 (minimal isometric) | — | 3 | 4 | — | 127 |
| 2 (tetrahedral) | 6 | 3 | 4 | — | 128 |
| 3 — | 3 | 3 | 4 | — | 129 |
| 4 — | — | 6 | 4 | 3 | 130 |
| 5 (full cubic) | 9 | 6 | 4 | 3 | 10-17 |

An exercise like that on page 126, using commas on a dodecahedron, will show you what happens when you examine a consistent

set of axes of fivefold rotation symmetry. Two fivefold axes entail all the rotational symmetry of the dodecahedron but none of its reflection planes. You have seen right-handed and left-handed solids with this symmetry on pages 39, 45, and 53. Adding a plane of reflection entails the full array of reflection planes of the dodecahedron.

The ornamented icosahedra below show that the full dodecahedral symmetry includes at a minimum the symmetry tabulated at 3 and is therefore isometric. Thus you see that not only all the regular solids but also all the Archimedean solids and their duals are isometric. But the prisms and antiprisms are not.

A pleasant way to explore the five symmetries tabulated on page 132 is to stellate the rhombic dodecahedron. Notice as you go that the first stellation, which puts four-sided pyramids over each parental face, can be regarded as a compound of three identical interpenetrating octahedra. The repeated octahedron is not regular: it might be derived by squashing a regular octahedron along one of its fourfold axes. You have already seen it among the solids shown on page 58.

In the top row of stellations, the first two are right-handed and left-handed forms of a solid having symmetry 1 of the earlier table. Although it arises from extending the faces of a rhombic dodecahedron, it is not strictly acceptable as a stellation because its faces have breaks, as the accompanying diagrams show. The third, however, is a true stellation, with symmetry 3.

The second row contains the rhombic dodecahedron, its first stel-

lation, and two more, all preserving the full cubic symmetry 5 of the parent. The last in this row is the dodecahedron's ultimate stellation: in any further extension of faces they never cut one another off again.

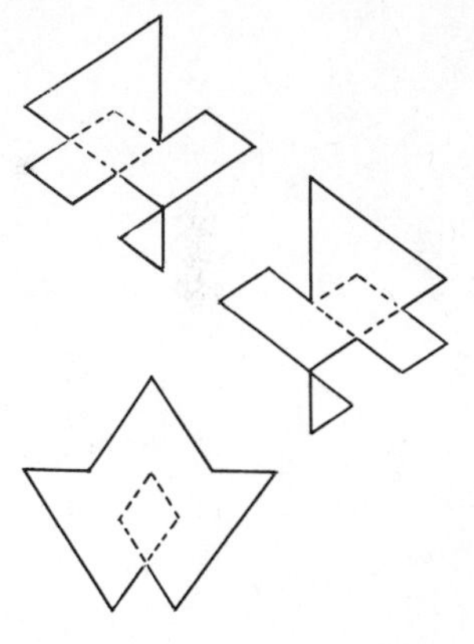

The lowest row shows a stellation with symmetry 2 and finally the left-handed and right-handed forms of an extension of the rhombic dodecahedron's faces that has symmetry 4. Again the last cannot be counted among true stellations because parts of its faces enclose no volume, but it has an interesting application that will appear later.

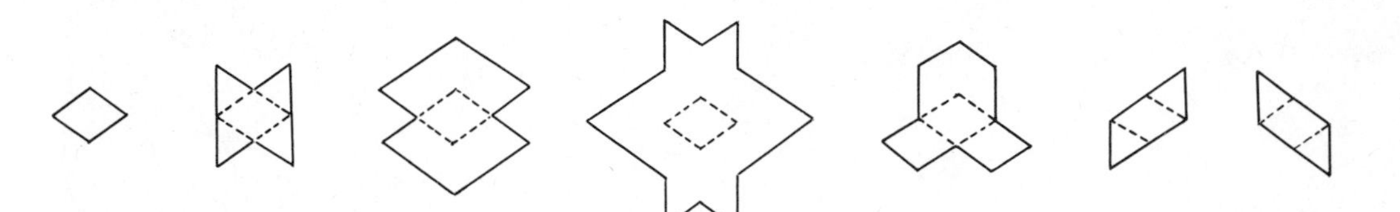

When you paint different colors on the faces of a solid, you raise two new questions the answers to which are sometimes fascinating. What happens to the symmetry of the solid when you take into account the coloring as well as the shape? What happens when you interchange the colors?

Each of the great stellated dodecahedra at the left clearly retains all the symmetry that it had before it was colored. Interchanging the colors on either solid transforms it into the other. If you take the two together, interchanging the colors and then turning the assembly upside down would leave its appearance unchanged.

Coloring four faces of an octahedron reduces its symmetry to that of a tetrahedron. Interchanging the colors then has the same effect as turning the octahedron through 90 degrees. Thus in the same sense that a tetrahedron is unchanged by interchanging its corners and faces (see p. 7), the colored octahedron is unchanged by interchanging its colors.

The regular nolid whose faces are four equilateral triangles (see p. 125) can be made in some especially interesting colored forms. Since the uncolored nolid has symmetry 4 of the table, it occurs in left-handed and right-handed forms. When it is made from sheets of cardboard whose opposite sides are differently colored, its six two-fold axes disappear and its three fourfold axes are reduced to two-fold axes. Thus it is degraded to symmetry 1. Interchanging the colors on either nolid leaves it unchanged.

The triangles forming the nolid can be colored instead in the two ways shown at the right: the front and back sides of the six segments are oppositely colored. Here each triangle looks the same whichever side you turn up, but a triangle of one sort is the mirror-image of a triangle of the other sort. You can distinguish four different nolids, each with symmetry 4. Interchanging the colors on all the nolids is equivalent to interchanging the upper pair with the lower pair. Reflecting the nolids is equivalent to interchanging the upper left with the lower right and the upper right with the lower left.

Colored designs on the faces of a rhombic dodecahedron can offer similar examples of the contributions of color to symmetry. Like the triangular nolids on the preceding page, all these decorated dodecahedra have symmetry 4. Again you can think of an interchange of colors as a potential *symmetry operation* like reflection through a plane. Can you arrange the decorated dodecahedra in an

array whose appearance will be unchanged if you reflect it, or if you interchange the colors, or both? Similarly, solids built of laminae whose edges are colored differently from their faces can illustrate interchanges of objects under symmetry operations that include both interchange of colors and reflection through a plane. Here, eight twisted pyramids are built on an octahedron.

The inside of a house is hard to visualize from its outside. The privilege of knowing what it looks like inside is reserved for its builders, its owners, its guests, and its burglars. The privilege of seeing what the faces on a solid look like from the inside is commonly reserved for those who make it.

But the ability to visualize an inside by looking at an outside is a generally useful skill and almost necessary for draftsmen, architects, and surgeons. Pictures can give a little help in cultivating the skill. On these pages you see the inside and outside of a great dodecahedron sectioned parallel to one of its planes of reflection symmetry. The pictures remind you that the inversion of viewpoint converts dimples to warts, mountain ranges to valleys, and waterspouts to whirlpools.

Does mankind look at the starry heavens from the inside or the outside? To that question you will find only a personal answer.

Like its outside, the inside of a solid can look very different from different points of view. These pages display two views of a solid that you have not seen before. Perhaps the pictures show enough for you to visualize what the solid looks like from the outside when it

is complete. Looking inside along an axis of fivefold rotation symmetry, you see a five-sided peak. Looking along a threefold axis you see a three-sided depression. The two views, added together in your mind, must accomplish your visualization, for they are all you have.

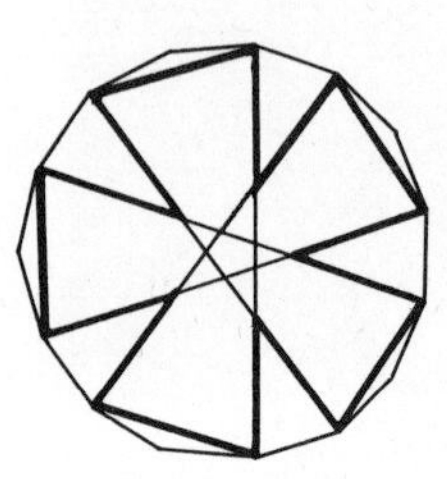

Whoever would make a model of this solid must visualize it in the following way. Its exterior faces are somewhat like regular starpentagons, but each face has two different kinds of sides, five of each kind, as the diagram shows. The sides of one kind meet one another to form exterior edges on the solid. At each of the remaining edges a side of the other kind joins a side of a regular pentagon. The twelve interior pentagonal faces intersect one another in somewhat the same way that the faces of the Kepler-Poinsot solids intersect. In order to construct the solid, your visualization must become exact enough to enable you to construct the exposed pieces of the pentagons, forming the sides and bottoms of the pits.

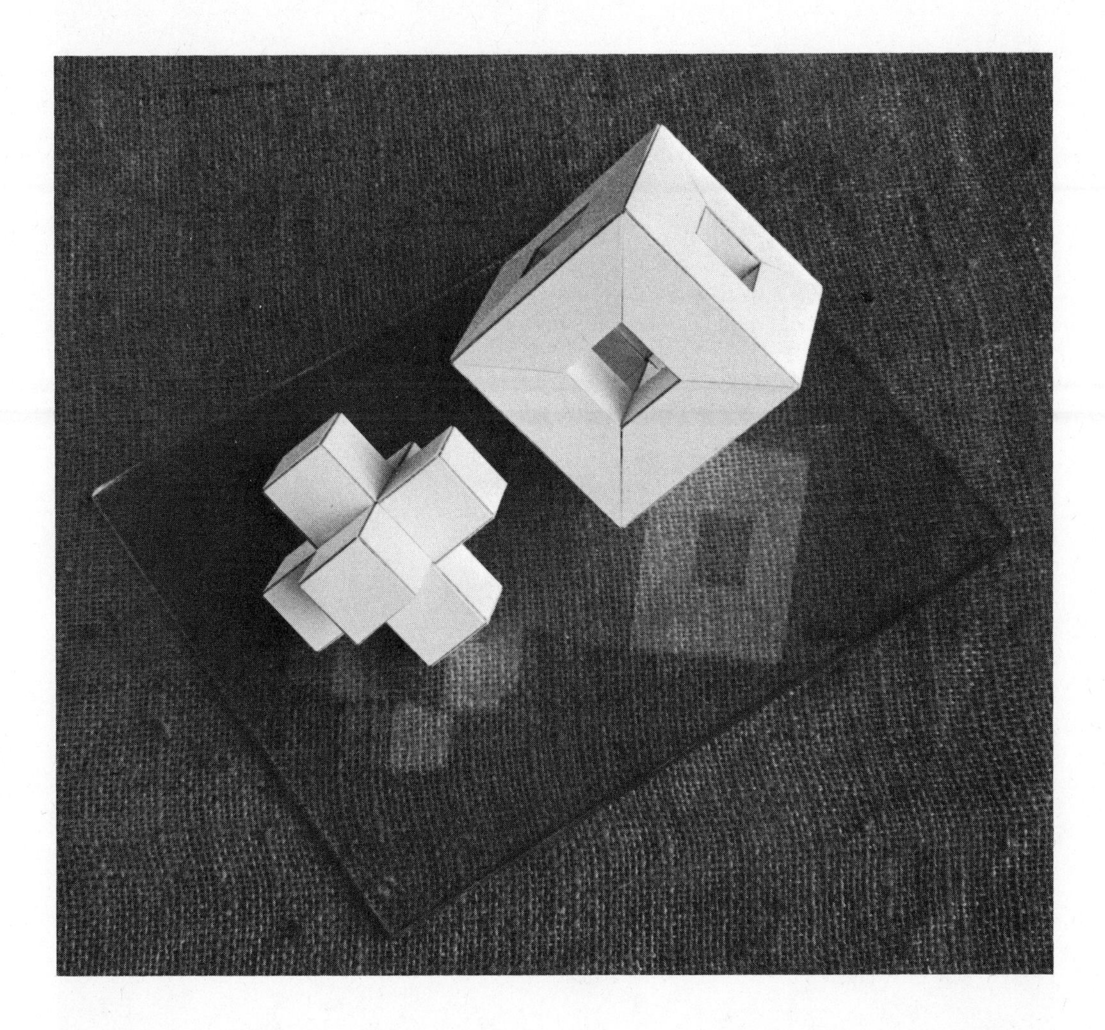

When you drill holes into a solid, you produce a new sort of inside, different from the inside of the original solid. Then you can make a solid whose faces are those of the new inside and look at the new solid from the outside. You might call that solid "the shape of the inside." Thus three square holes drilled through a cube from face to face give to the cube an inside that looks like three long square prisms intersecting one another. Since the drilled cube and its inside added together reproduce the original cube, you can think of the two solids as "supplements relative to a cube."

Drilling three square holes through a cube again, with the drill turned at 45 degrees to the edges, you produce two supplements that are more interesting. Again, of course, you can visualize the shape of the inside as three interpenetrating square prisms, because the drill is square and the holes are straight, but the shape of the supplement looks very different from that shown on page 147.

The new inside of the cube has shallow three-sided pyramids projecting into it, and its supplement has corresponding dimples. When you remove the pyramids from the inside, you add them to the supplement, filling its dimples and producing a cuboctahedron.

A drill whose cross-section has threefold symmetry, forced to the center of an isometric solid in one direction along each of its threefold axes, can yield quite unexpected pairs of supplements. A regular octahedron supplements the pierced tetrahedron. The pierced truncated tetrahedron, with its inside trimmed of projecting pyramids, presents an especially attractive supplement. Neither the pierced cube nor its supplement suggest immediately their cubic origins.

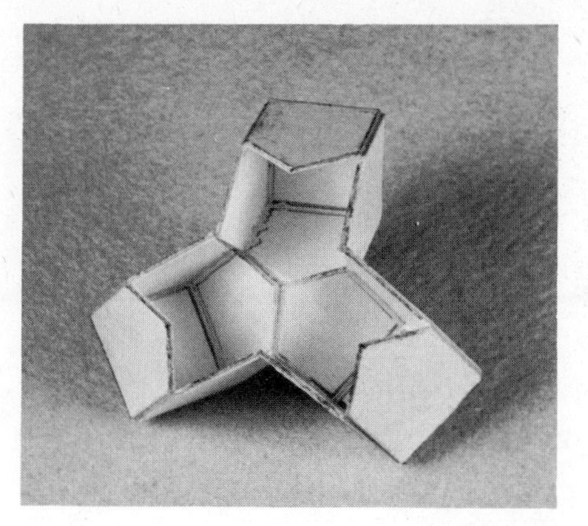

Looking at the *inside of an inside* can also assist visualization: it may make clearer what you would see if you stood in the middle of a pierced solid.

The solid whose inside appears on the previous page, in views along a threefold and a fourfold axis, is the supplement to the solid produced by drilling six square holes through a great rhombicuboctahedron along its axes of twofold rotation symmetry. The holes are just large enough to remove completely the twelve square faces of the solid. After the drilling, six square pyramids and eight triangular pyramids project into the interior, as you saw by studying the insides of the insides on the previous page. When you trim off the pyramids, the solid and its supplement become what you see above. The trimmed pyramids contribute six square faces and eight triangular faces to the supplement.

And now it seems natural to trim the projecting cubes from the supplement: the remaining solid is a small rhombicuboctahedron. Thus you discover the curious fact that a small rhombicuboctahedral interior can be connected to a great rhombicuboctahedral exterior by cubical holes that obliterate twelve faces of each. By analogous operations on a great rhombicosidodecahedron you can construct an interior shaped as a small rhombicosidodecahedron and connected to the outside by thirty holes shaped as square prisms. In that case, however, the holes are not cubical.

As everyone knows, rectangular boxes that are all of the same sort can be stacked together indefinitely, leaving no open spaces between them. Solids with many other shapes will also fill space densely: you can imagine distorting the boxes so that they are not rectangular, as the diagram suggests. If you require that the solids be isometric, however, you place a stringent limitation on your search; the only isometric rectangular box is the cube, and distorted cubes are not isometric.

The impulse to make a space-filling box isometric lies in the fact that free space seems indifferent to direction. Space offers no preferred axis along which to align the length of a long rectangular box. Recognizing this indifference, mathematicians often refer their three-dimensional calculations to three axes that are mutually perpendicular and indistinguishable, named *Cartesian axes* for the seventeenth century philosopher René Descartes. The three twofold axes of an isometric object have those two Cartesian properties: they are mutually perpendicular, and the accompanying threefold axes make them indistinguishable.

Clearly the cube provides the only *regular* shape whose repetition will fill space densely. But since all the Archimedean solids and their duals are isometric (see p. 133), they form an inviting realm in which

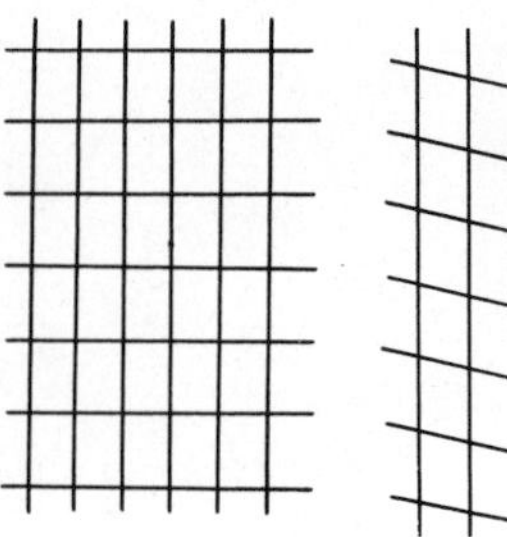

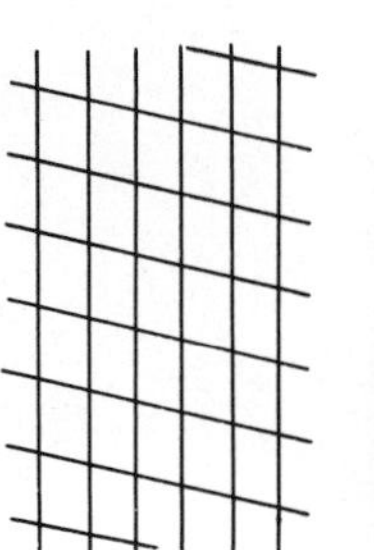

to search further. The search turns up two. Rhombic dodecahedra pack densely together, each sharing its faces with twelve of its fellows. And truncated octahedra pack to leave no space between them by sharing their hexagonal faces with eight immediate neighbors and their square faces with six neighbors somewhat more distant.

Extending the search for isometric space-filling solids beyond the Archimedean solids and their duals, you can find some ways of systematizing the effort, but none will provide a complete set. A surprising and little-known example of such a solid is the first stellation of the rhombic dodecahedron, which you have already seen on page 134. Some of the edges of that solid appear in six square crosses, and they can be brought up against similar crosses on six fellows. Then eight other fellows fit exactly against that assembly of seven along its axes of threefold rotation symmetry. More fellows can be added to extend the arrangement indefinitely.

The six square crosses formed by the edges of the stella octangula invite the treatment given on the opposite page to the stellated rhombic dodecahedron. When six solids are placed against one in this way, the remaining depressions do not accommodate a stella octangula, but they do accept regular octahedra. Thus you find a case in which two isometric solids combine to fill space. In fact, noticing that each stella octangula can be made by fastening eight tetrahedra to the faces of an octahedron, you see that space can be filled by intercalating *two regular solids,* the octahedron and the tetrahedron.

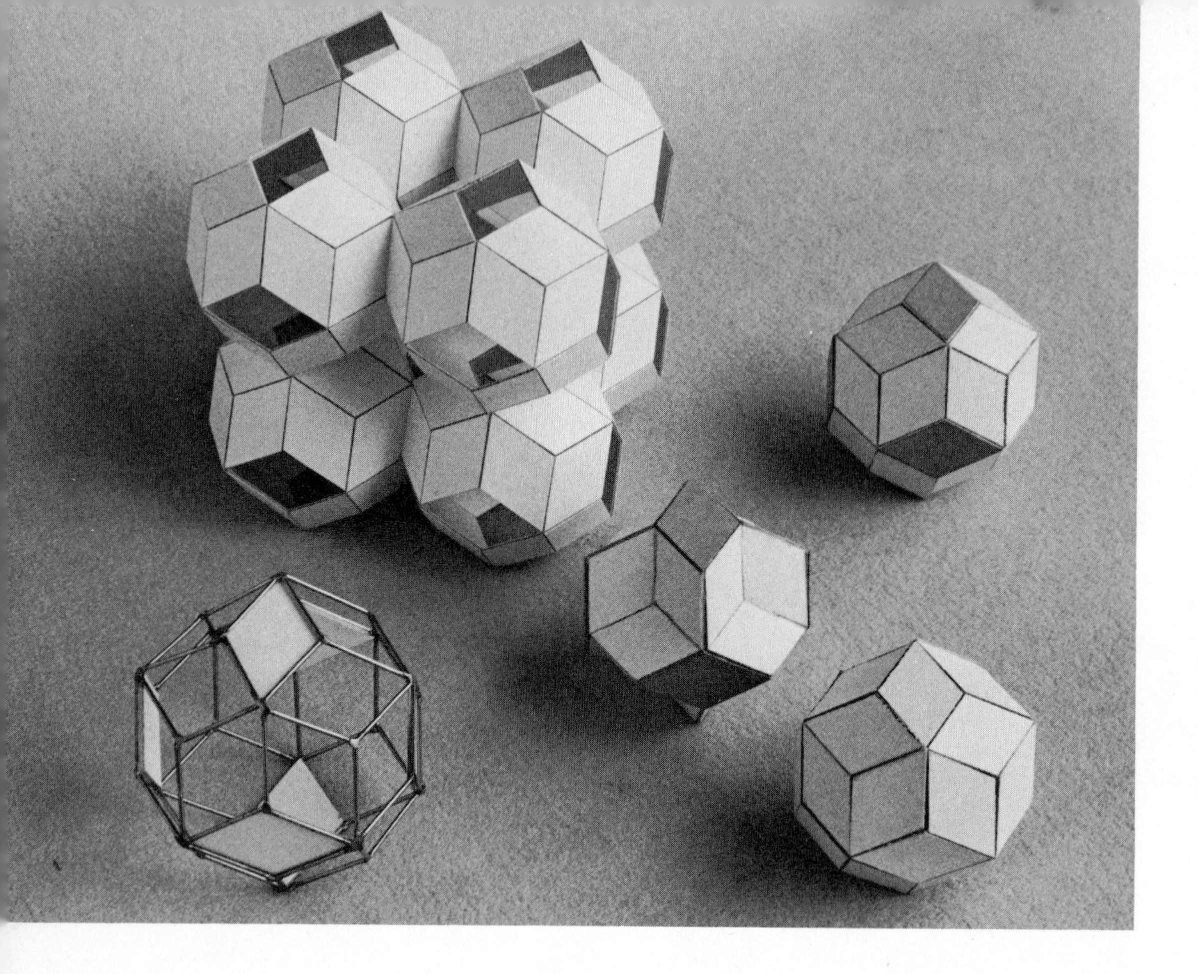

Since the rhombic triacontahedron is an Archimedean dual, it is isometric. Six of its faces can be juxtaposed against faces of its fellows, and more fellows can join them, to lie in an extended arrangement where the solids occupy the positions of stacked cubes. But the arrangement leaves open spaces between the solids. Exploring the character of those open spaces requires visualization of the same sort that you used earlier to examine the holes in solids. In this array the holes turn out to have a shape made of the same rhombuses as the solids themselves. But the shape of the holes is concave at eight places along threefold axes where the solids are convex. Thus the array and its supplement fill space with two sorts of isometric solids in equal numbers, and both sorts separately occupy the positions of stacked cubes.

The last of the stellations of the rhombic dodecahedron shown on page 135 has the shape of the dodecahedron's first stellation with parts of the facial extensions obliterated. This solid therefore fits its fellows in the same surprising way that the first stellation fits (see p. 156), and the obliterations open up straight-running holes through the array. Thus you discover the strange fact that you can fill space with rhombic dodecahedra in a different arrangement from that on page 154 by supplementing them with rods of triangular cross-section that extend indefinitely. Since the stacked stellations come in right-handed and left-handed forms, the grid of rods could be chosen in either of those ways, one the mirror image of the other, without disturbing the supplementary array of rhombic dodecahedra.

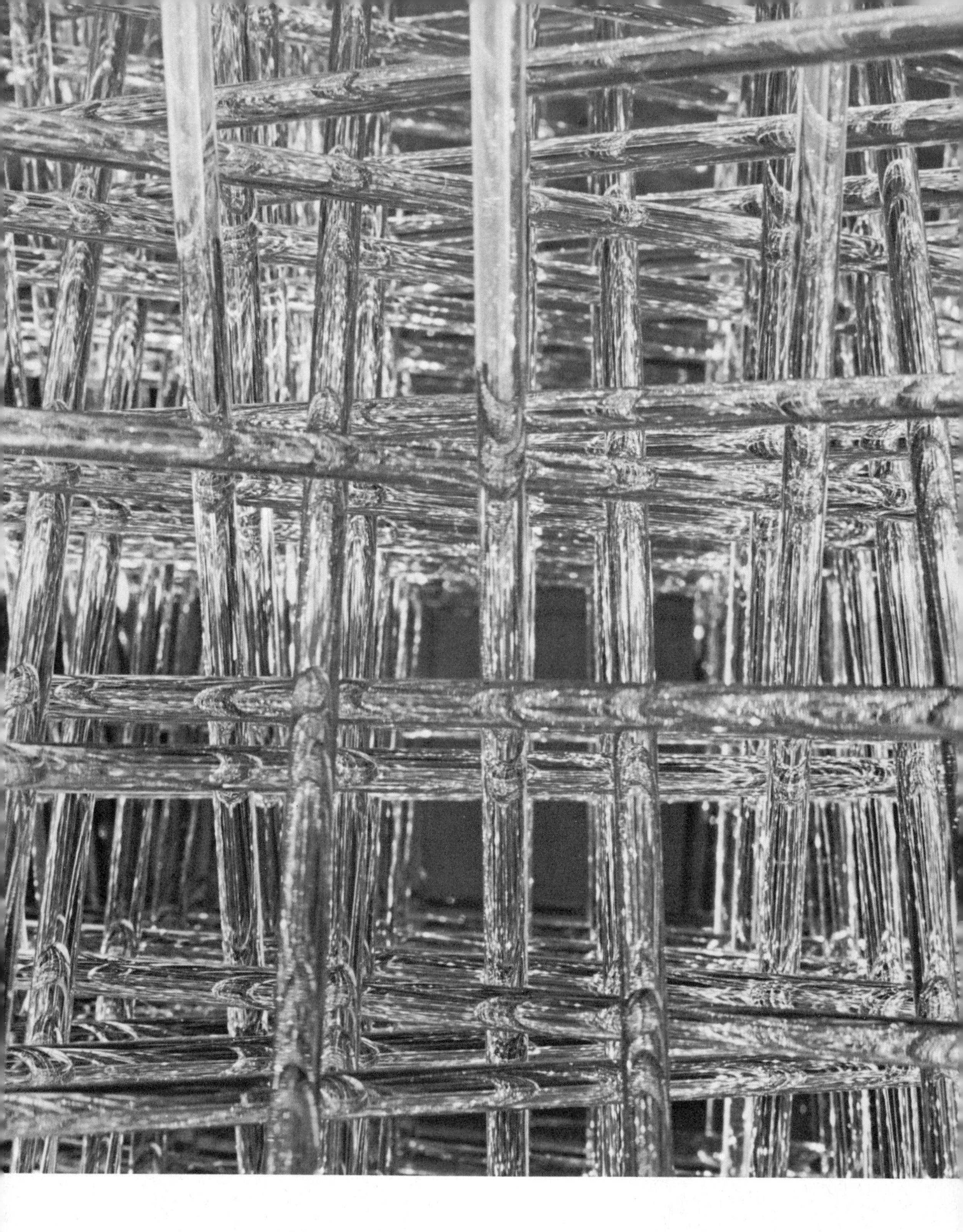

Look for a moment at this discovery from a different point of view. The rods that supplement the dodecahedra mark the course of straight paths through space, which thread past one another without intersecting. They sort themselves into four families; within each family the paths are parallel and regularly spaced. Each family is parallel to one of the threefold axes of rotation symmetry of the parental dodecahedra, and the four families together have isometric symmetry.

A cube fathers a comparable array of paths with a simpler structure. Here three nonintersecting families of paths take courses parallel to the cube's fourfold axes. Again the whole array is isometric, but now without distinguishable right- and left-handed forms.

Looking at space-filling solids, you saw several examples of arrays in which the repeated solids occupy the positions of stacked cubes. A convenient way to describe those arrangements is to specify what they all have in common, the locations of the centers of the solids. Points in space represent those centers; when the arrayed solids are

all alike, the arrangement of points is called a **lattice**. In particular, the centers of stacked cubes occupy a *simple cubic lattice.*

Lattices are often portrayed by centering little spheres on the lattice points and connecting the spheres by sticks. Here little octahedra replace the spheres and emphasize that each lattice point has six nearest-neighboring lattice points located on axes of fourfold rotation symmetry of the array.

Turning to the stacked rhombic dodecahedra (see p. 154), you find that their centers occupy a different lattice, in which each point has twelve nearest neighbors located on twofold axes of the array. The arrangement is one of those that identical spherical balls can take in order to occupy a minimum of space. Since it is the only one of those arrangements that is isometric, it is often called *cubic closest packing.*

A third lattice appears in the arrangement of centers of the stacked truncated octahedra (see p. 155). The eight nearest neighbors of each lattice point fall on axes of threefold rotation symmetry of the array. The centers of the six remaining solids making contact with each solid occupy lattice points that are "next-nearest neighbors" in the same lattice. Here you could draw a cube centered on any lattice point and find the nearest neighbors of that point at the eight corners of the cube. For that reason the arrangement of points is called a *body-centered cubic lattice.*

Notice that the stacking of stellated rhombic dodecahedra locates their centers also on a body-centered cubic lattice. Thus a lattice does not prescribe uniquely a shape that belongs with it. But studying the packing of cubes, rhombic dodecahedra, and truncated octahedra has brought to light three arrangements of points in space that are called *homogeneous* because each point is identically surrounded by its fellows when the lattice extends indefinitely. Standing at any point in such a lattice and looking in any particular direction, you see the same environment. These three are the only homogeneous lattices that are isometric.

In the stack of rhombic triacontahedra previously examined, the solids occupy a simple cubic lattice. They leave holes between them that might be said to "occupy" another simple cubic lattice interlacing the first. Indeed, by placing in those holes the supplement shown on page 158, the second lattice would become occupied in the full sense of the word. Notice that the two interlacing lattices seem to form a single body-centered cubic lattice. But standing on a point in one of the component lattices you would not see the same environment that you would see from a point in the other component, if you take into account the differences in the solid shapes that are symbolized by the points. It is well to speak here of *two interpenetrating simple cubic lattices.*

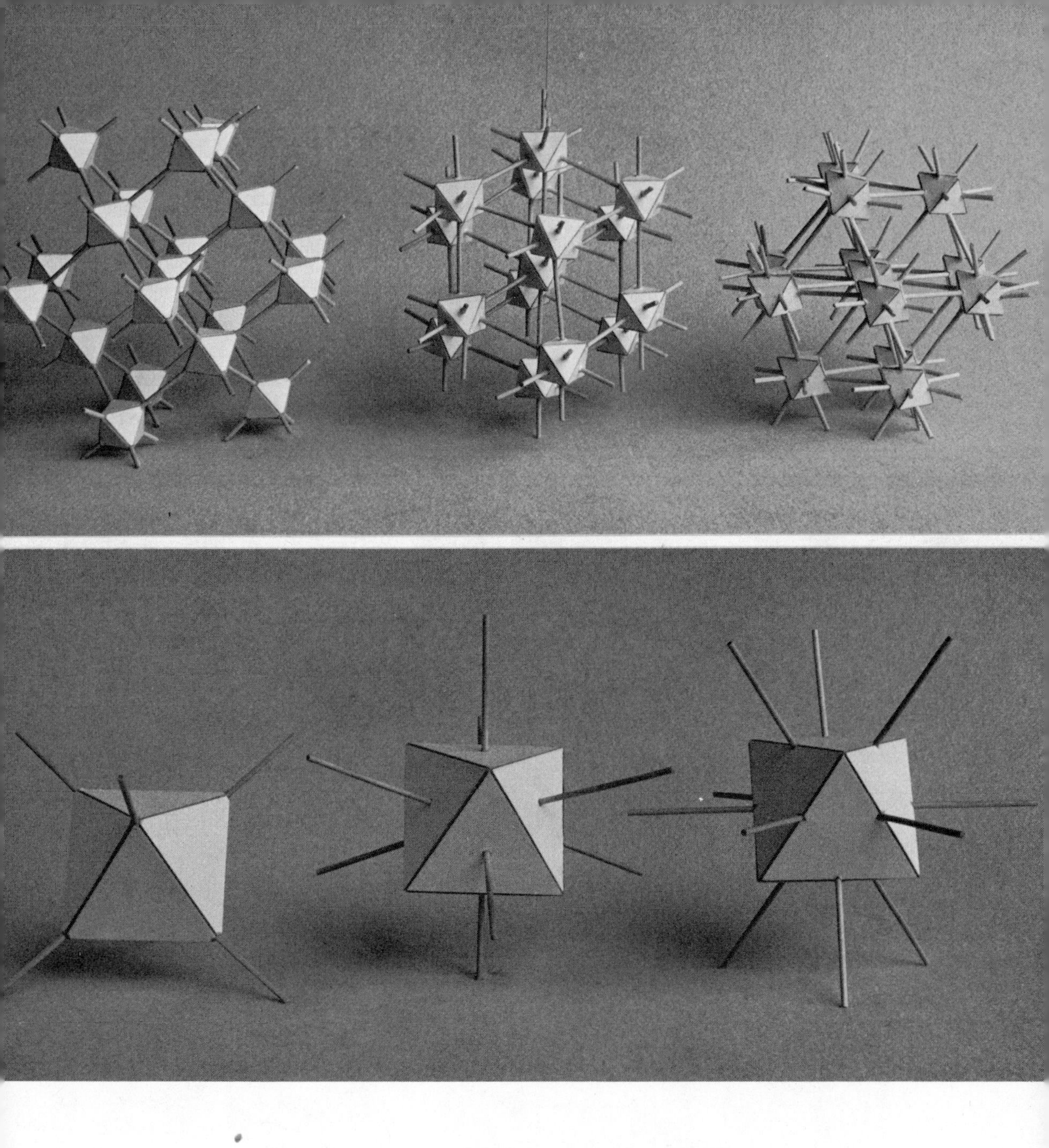

The three homogeneous isometric lattices correlate closely with the three kinds of rotation symmetry of the cube and the octahedron. The fivefold symmetry characteristic of the dodecahedron and the icosahedron cannot appear in lattices.

Since the three isometric lattices are repetitive arrays of points, they can be portrayed by showing how only a few of the points are related to one another: the portrayal can be chosen so that repeating it will build the whole lattice. It is often advantageous to choose the portrayal so that it has the symmetry of the lattice when it stands alone, without repetition. The three isometric lattices have more than the minimal isometric symmetry; they have the full symmetry of the cube. Thus the cubical arrangements shown below, taken from left to right, are appropriate to the simple cubic, body-centered cubic, and closest-packing lattices. Notice in the last that the arrangement makes suitable the name *face-centered cubic* for the lattice.

Because repeating them will build the lattices, these portrayals are called *unit cells.* How many lattice points does each unit cell contain? The repetition places the faces of the cells in contact, and therefore a lattice point on a face of a cell is shared with another cell, and a point at a corner is shared with seven other cells. In other words, the eight portrayals at corners of a cell account for one lattice point, and the six portrayals on faces account for three. The cells shown below account for one, two, and four lattice points, respectively.

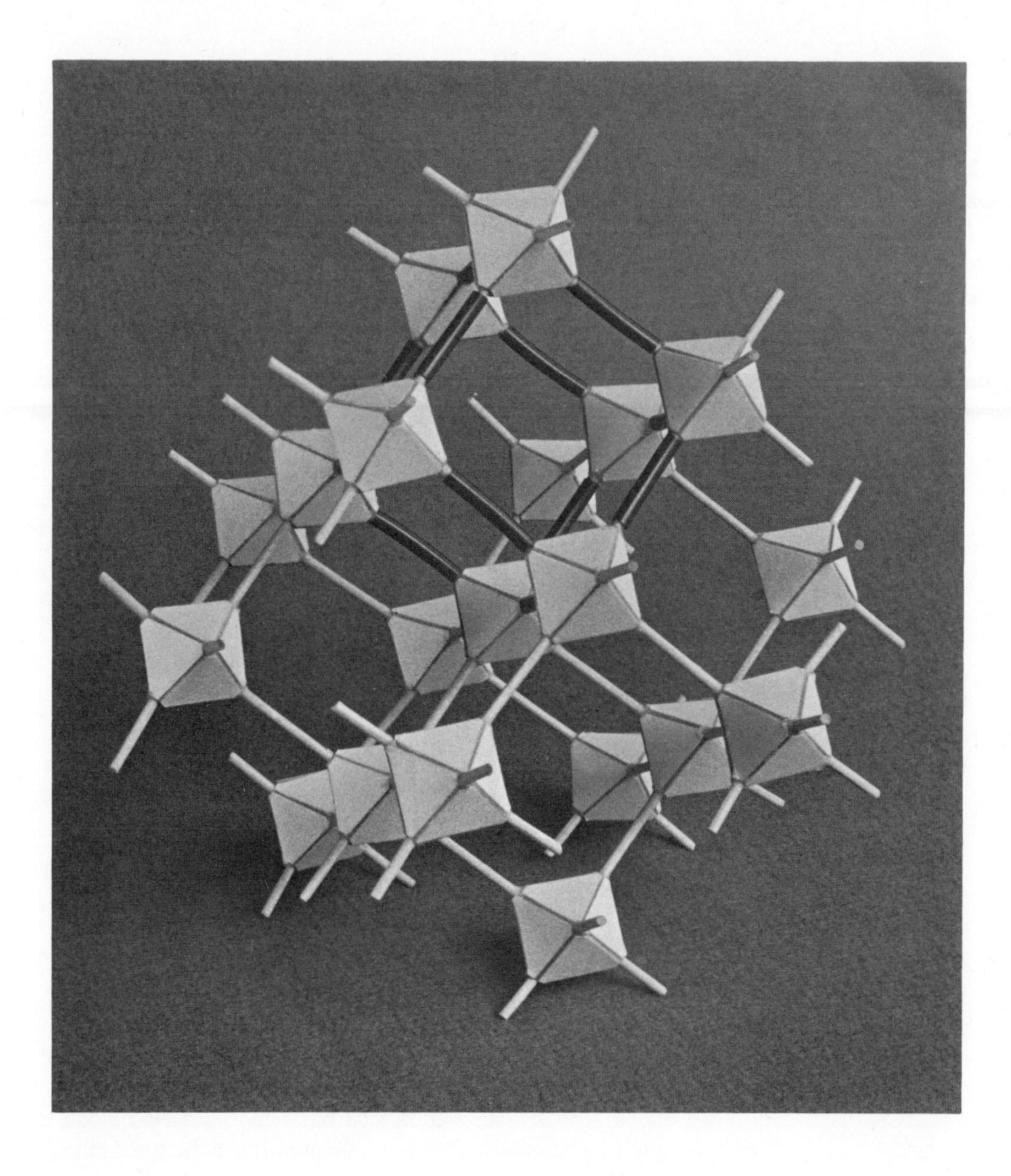

Outlining in black one of the unit cells of the simple cubic lattice may help to make these ideas graphic.

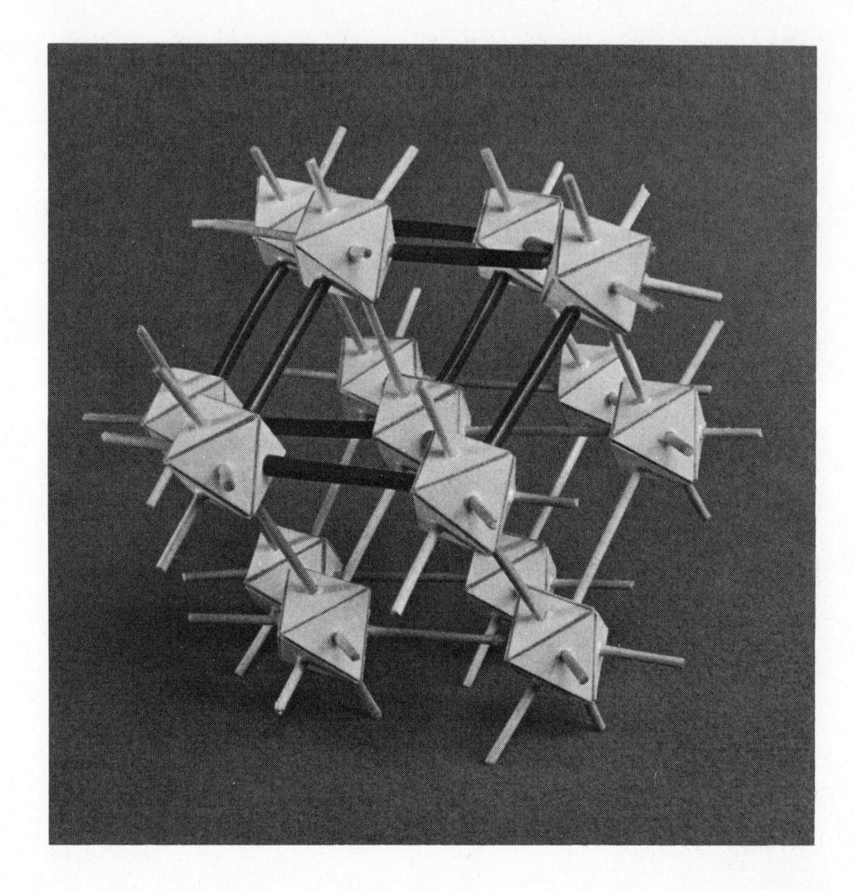

Since all the points on a lattice are equivalent, it is possible to choose for each lattice a unit cell that accounts for just one point. The cubical cell for the simple cubic lattice already has that property. For the body-centered cubic lattice, a cell accounting for a single point is outlined above: it looks like a cube that has been squashed along one threefold axis. Correspondingly, the cell outlined on the facing page, looking like a cube stretched along a threefold axis, accounts for one point in the face-centered cubic lattice. To achieve their virtue of economy these one-point cells, called *primitive cells,* sacrifice the portrayal of symmetry.

In this arrangement of truncated icosahedra, each makes contact with four others whose centers can be imagined to lie at the corners of a tetrahedron surrounding it. The centers can be divided into two groups for which the surrounding tetrahedra are oppositely cocked, and therefore the arrangement of centers cannot properly be called a lattice. To be sure, each group of points taken separately forms a face-centered cubic lattice. But together the points form a *structure,* often called "the diamond structure" because it locates the centers of the atoms of carbon that compose a crystal of diamond.

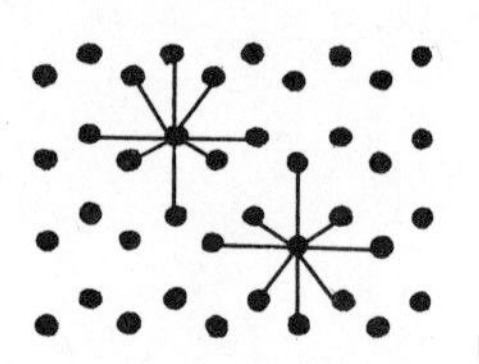

Again for the diamond structure a space-filling shape can be found in which the centers of the repeated shapes fall on the points of the structure. The shape resembles a truncated tetrahedron, but with shallow trigonal pyramids replacing the truncated corners. The same shape accommodates the two distinguishable groups of points in the structure by cocking oppositely in the assembly of solids.

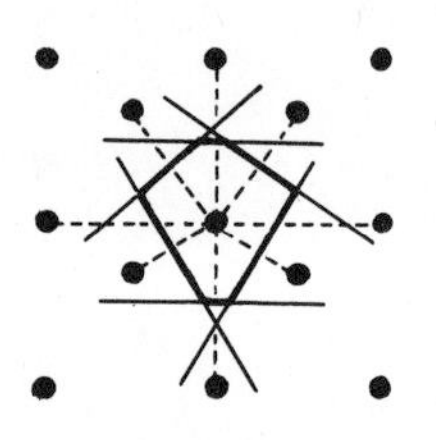

There is a recipe for finding the space-filling shapes appropriate to any structure of points. (1) Draw lines connecting a point to all its neighbors. (2) Draw planes perpendicular to those lines and bisecting them. (3) Pick out the shape that is enclosed by all these intersecting planes and is not cut by any of them. The diagrams show a comparable construction in two dimensions for a structure whose points fall in two different groups.

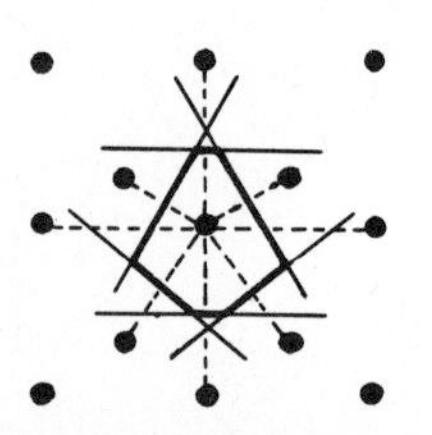

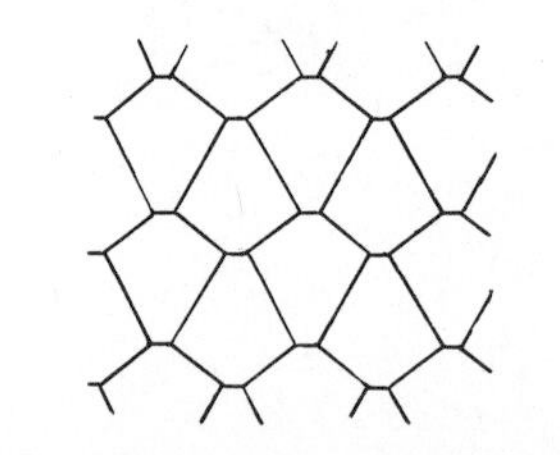

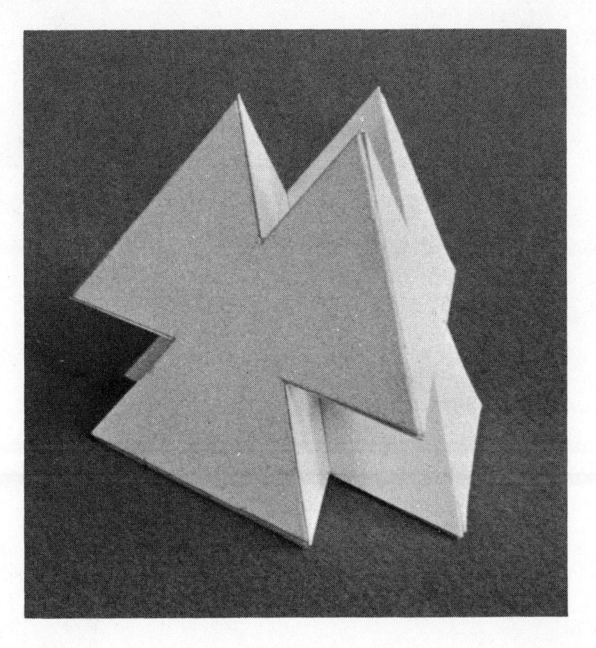

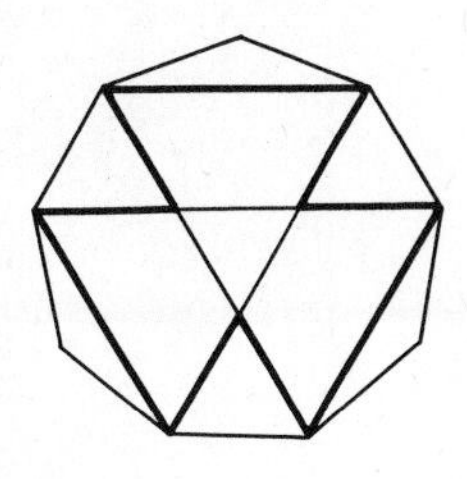

The construction of the solid pictured on page 146, with faces made of sectored pentagons, may spur you to perform a comparable experiment with sectored triangles. When four such figures are disposed like the faces of a tetrahedron, their intersecting edges can join with the edges of four large intersecting triangles. Again the product is a closed solid, which is somewhat like a semiregular solid because a triangle and a sectored triangle join at each interior edge, and each triangle meets two sectored triangles at each corner.

But now it is instructive to look at this solid differently: the solid decomposes into six smaller tetrahedra that intersect one another in a new way. Unlike the centers of the two interpenetrating tetrahedra in the stella octangula, the centers of these six tetrahedra do not fall at a common point. They are distributed like the six corners of a regular octahedron.

The idea of *partly interpenetrating solids,* applied to the simpler solids shown early in this book, can suggest some interesting constructions. Some of those appearing on the next few pages make use of such geometric properties of the solids as the shapes of their cross-sections (see pp. 14–25). Others take advantage of their symmetries to dictate successful juxtapositions into rings and chains.

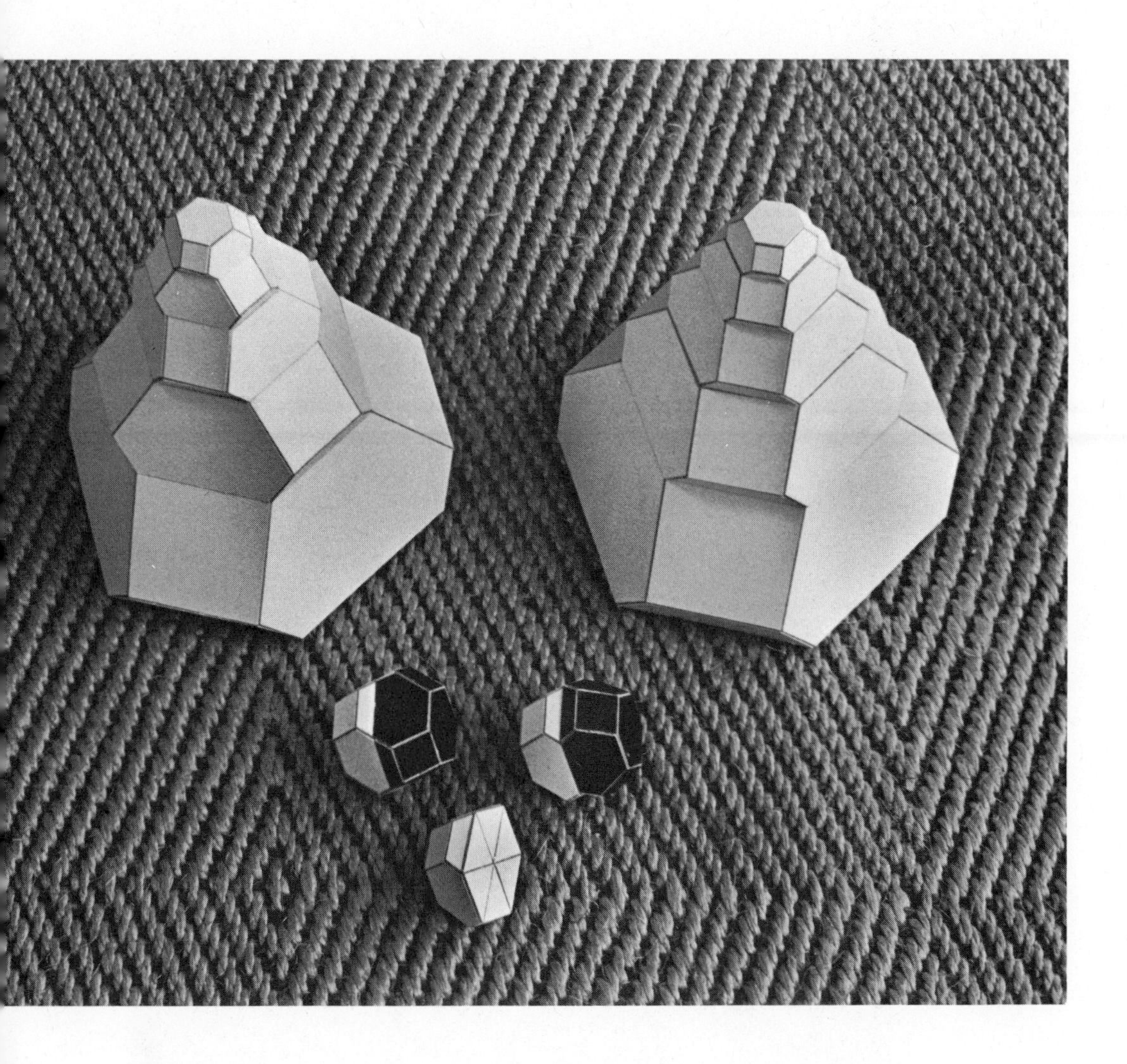

The square cross-section of a cuboctahedron can be placed upon the square face of a larger cuboctahedron. Repetition of that operation constructs a tapering tower. Since the sizes of the successive cuboctahedra fall in a geometric series, the height of the tower is finite, even if you add cuboctahedra indefinitely. Similar towers of partly interpenetrating truncated octahedra can be built in two ways.

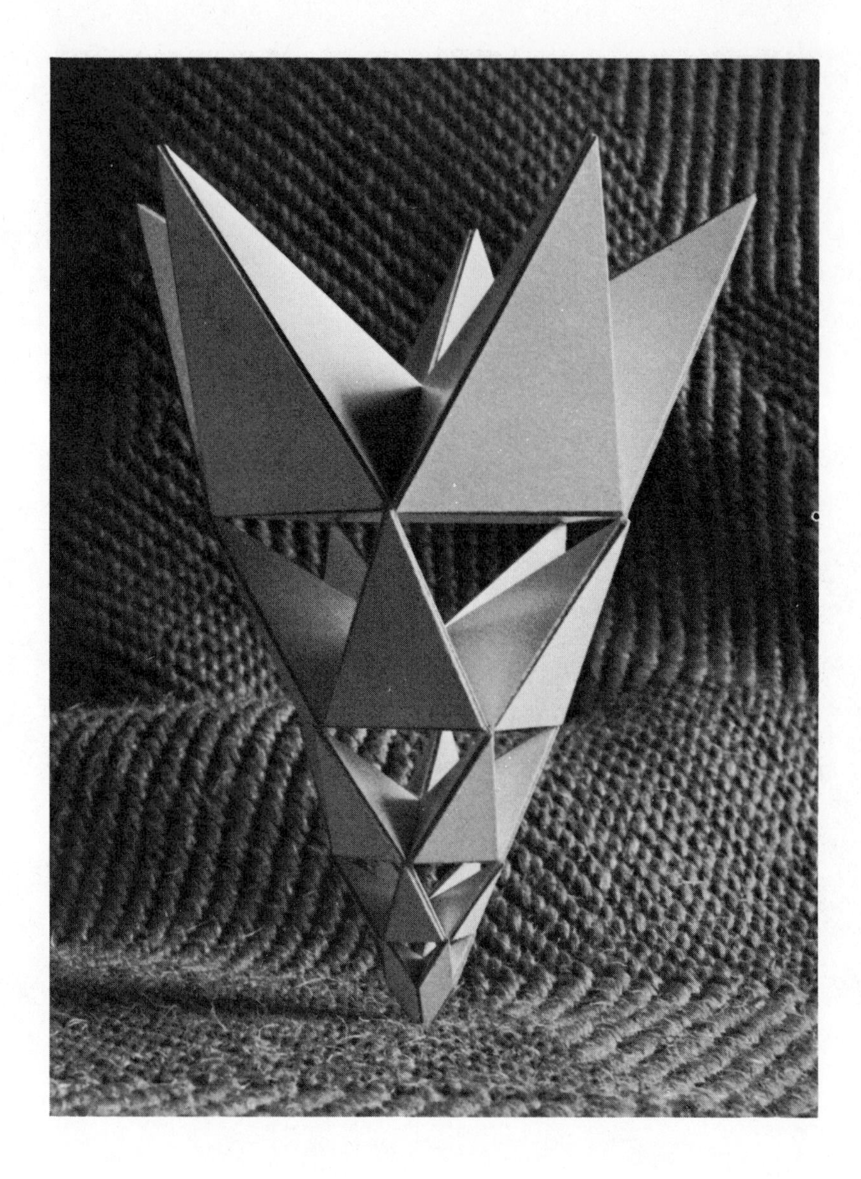

The accumulation of sectioned great stellated dodecahedra is reminiscent of botanical growth forms. But successive stages in the growth of living things do not commonly fall in a geometric series. Sir D'Arcy Wentworth Thompson has given a classic description of their serial habits in his book *On Growth and Form.*

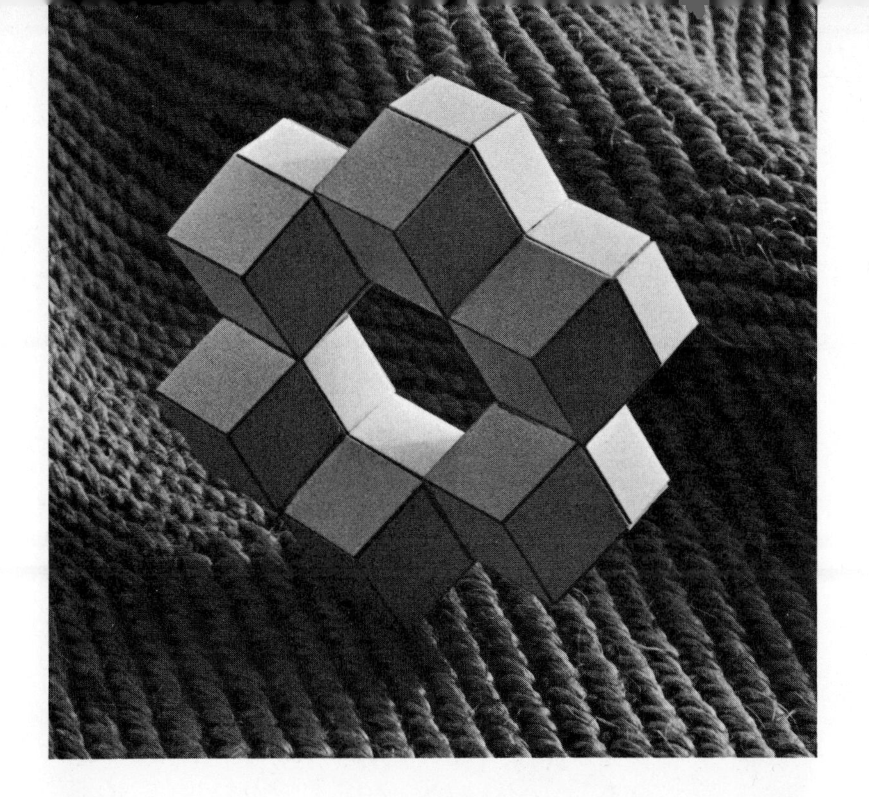

The juxtaposition of six rhombic dodecahedra in a ring follows directly from the packing of those solids to fill space (see p. 163). But the possibility of juxtaposing octahedra to form four rings fused together isometrically is less immediately obvious.

Twelve rhombic dodecahedra can be assembled into a puckered ring in which each partly interpenetrates its two neighbors along their axes of fourfold rotation symmetry.

When rhombic dodecahedra partly interpenetrate along threefold axes, they can be turned by 60 degrees with respect to their neighbors to form a similar puckered ring.

Rings linked through one another in succession form the familiar chain, which cannot be taken apart completely without cutting all the rings. More complicated ways of linking rings inseparably arise from examining nolids. The edges of the two regular nolids made of four equilateral triangles (see p. 125) define the course of four triangular rings, linking without intersecting and forming an isometric assembly. A similar regular nolid made of six pentagons, and another of ten triangles, define isometric assemblies of six pentagonal and ten triangular rings. In each case the assembly can be made in a right-handed or a left-handed form, and even if the rings were made of a flexible material such as rope, one form could not be manipulated into the other form without cutting the rings. And each ring is linked with every other.

The nolid made of three regular hexagons, though isometric, is not regular because it has two different sorts of corners. But the rings defined by its edges interlock in an interesting way. The rings are inseparable, but no two are linked together; hence cutting any one frees the other two.

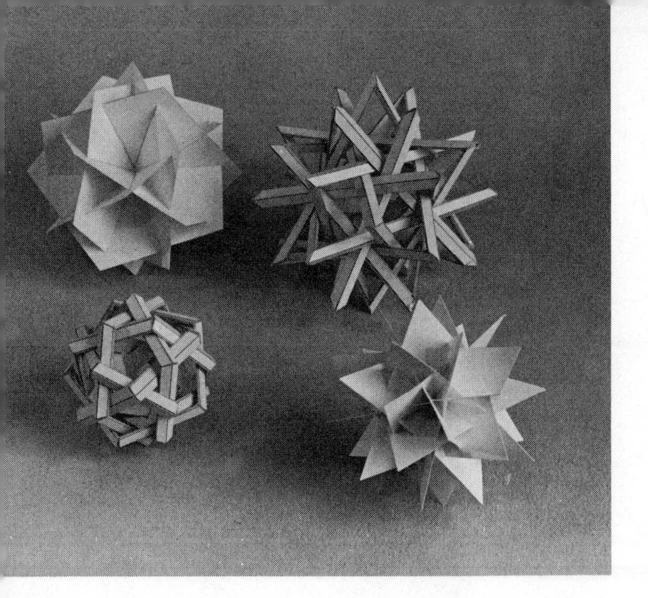

If the hexagons are turned into circles and the assembly is laid flat, it forms the configuration sometimes called the Borromean rings (from its appearance on the arms of the Italian family, Borromeo) and sometimes the Ballantine (from its use as the trademark of an American brewery). In its flat form the Ballantine appears to offer distinguishable right-handed and left-handed options. But the parental nolid made of three hexagons has symmetry 3, whose planes of reflection exclude such options. If you make the loops of flexible material, you will find that in fact one form can be manipulated into the other without cutting the loops.

Everyone ties knots and comes to know when he unties them again that the idea of a knot is a complicated idea, much more complicated than the idea of a ring. To a mathematician a *knot* is a closed curve in three dimensions that cannot be turned into a *ring* without cutting it. If a ring is made of something flexible, it can be distorted to look like any other ring; in other words there is only one kind of ring. But there are many kinds of knots and no wholly satisfactory way to classify and define them. If you flatten a knot and count how many times one part of it crosses another, however, you have begun an analysis of it. Three crossings is the smallest number that you will ever find, and you can think of the **trefoil knot** as the simplest possible knot.

Now make an excursion from point to neighboring point through a simple cubic lattice whose course forms a trefoil knot. Placing cubes on the lattice points traversed, and fusing them face to face, you can form the shape shown above. It is a solid with one axis of threefold rotation symmetry and three twofold axes perpendicular to the threefold axis.

Remembering that the body-centered cubic lattice can be obtained by squashing the simple cubic lattice along a threefold axis (see p. 170), you can convert the previous excursion into a body-centered cubic excursion. Then you can fuse regular octahedra into a trefoil knot that has the same symmetry as the knot made of cubes.

Since the trefoil knot has no planes of reflection symmetry, it can be tied in right-handed and left-handed forms. If you make four replicas of either form and fit them together so that their threefold axes fall along those of a cube, you will find that they leave a space in the center that exactly accommodates a little cube. The dense assembly has four threefold and three twofold axes of rotation symmetry. When the ingredients of the assembly are fused to form a single solid, in which the outlines of the knots are obliterated, the twofold axes become fourfold, and six new twofold axes appear. In other words, the assembly with symmetry 1 becomes a solid with symmetry 4.

The isometric assembly of four trefoil knots suggests a fascinating question. Is it possible to make a single knot that is isometric? As far as this writer knows, nobody has answered that question.

The solids pictured in this book were all assembled from individually cut flat faces. Suitable cardboard is of the sort used for mounting photographs, and suitable glue is a "white glue" of the sort exemplified by "Elmer's Glue-all." This method of construction accents the edges of the solids pleasantly and unobtrusively; the accented edges assist the faces in clarifying the finished structure.

The faces to be assembled are best cut by mounted shears. The most useful models of this tool have a movable barrier or "fence," which can be clamped parallel to the cutting edge at any distance from it. Cutting a large number of congruent faces is best carried out by first cutting strips of uniform width, and then sliding a strip along the cutting board with one edge against a guide that establishes the desired cutting angle. Equilateral triangles, for example, are made from strips whose width is the altitude of the desired triangle. The guide is a draftsman's 30-60-90 triangle, or a corresponding piece of cardboard. After one cut, the strip is turned over and the vertex is registered by eye over the lower blade of the shears to

establish the position of the next cut. Repetition of this operation over the length of the strip produces a collection of congruent triangles.

Regular hexagons are best cut by first cutting equilateral triangles and then truncating their vertices with the aid of a fence set back from the cutting edge by two thirds of the altitude of the triangles. To cut squares, use the fence to cut strips and leave it in the same position to cut the squares from the strips. Cutting other polygons requires a guide made of heavier cardboard, such as an illustration board, with two edges meeting at the desired angle.

The photographs on this page and the next show these operations in progress to produce faces for rhombic dodecahedra. The acute angle of the rhombus is 70 degrees 32 minutes; but the shape of the desired polygon can be found, without using a protractor, by the method of construction described on page 51. After a cut, the strip is moved a distance equal to its own width, gauged by holding another strip against the cutting edge.

The faces are usually best assembled first in pairs. A little white glue is squeezed out of its squeeze-bottle to form a small pool on a 3" x 5" file card. A toothpick is dipped in the glue and used to smear one side of a polygonal face. The side of the face to be attached is pushed against the glue-bearing side and slid along it to distribute the glue. The pair is then placed on a table with the glued joint standing up. More such joints are made, and after a few minutes the first-made joint has set sufficiently to permit further work with the pair. For several hours the joint remains sufficiently flexible to permit adjustment of the dihedral angle.

In making an octahedron, the appropriate sides of one such pair of triangles are smeared with glue, and the two pairs to be joined

are slid toward each other along a flat surface, to form a square pyramid. Two square pyramids are later joined at their bases to complete the octahedron. In making a rhombic dodecahedron it is best to glue one more rhombus to a pair of rhombuses, providing three rhombuses meeting at a threefold axial corner and thus fixing their dihedral angles. Four such triples can then be joined successively to complete the solid.

In making more complicated solids, both ease of assembly and rigidity of the product are promoted by assembling the external parts on a simpler solid base whenever that is possible. The stella octangula can be made of four trigonal pyramids attached to faces of a tetrahedron, and the great stellated dodecahedron can be made of twenty pyramids on the faces of an icosahedron.

Wire models are best made by using cardboard models as jigs to hold the wires, which are appropriately of no. 16 B&S gauge tinned copper. Each piece of wire, cut to the length of an edge, is fastened to it with white glue extending a little short of the corners. The corners are soldered, and trimmed if necessary with a flat file. The assembly is soaked in water long enough to soften the cardboard, and the faces are pushed inward in succession and removed.

**Bibliography.** Most of the books listed below provide further bibliographies.

Critchlow, K. *Order in Space.* New York: Viking Press, 1970.

Cundy, H. M., and A. P. Rollett. *Mathematical Models.* 2nd. ed. London: Oxford Univ. Press, 1961.

Heath, T. L. *A Manual of Greek Mathematics.* New York: Dover Publications, 1963.

Klein, F. *Lectures on the Icosahedron.* New York: Dover Publications, 1956.

Lines, L. *Solid Geometry.* New York: Dover Publications, 1965.

Thompson, D'A. W. *On Growth and Form.* Abridgement by J. T. Bonner. London: Cambridge Univ. Press, 1961.

Wells, A. F. *The Third Dimension in Chemistry.* London: Oxford Univ. Press, 1956.

Wenninger, M. J. *Polyhedron Models.* London: Cambridge Univ. Press, 1971.

Weyl, H. *Symmetry.* Princeton: Princeton Univ. Press, 1952.

**Index.** Antiprisms: convex semiregular, 61; convex trapezohedra dual to, 62; star-polygonal semiregular, 76ff, 80; duals to star-polygonal, 81
Archimedean duals, 50ff; circumscribed to spheres, 51; suitability for dice, 53
Archimedes, 46
Axes, Cartesian, 154
Axes of rotational symmetry, 10ff, 20, 34ff

Ballantine, 183
Borromean rings, 183

Cartesian axes, 154
Cauchy, Augustin, 83, 93
Cells, unit, 168; primitive, 170
Color symmetry, 137ff
Compound solids, *see* Interpenetrating solids
Consistent symmetry operations, 127, 132
Convex solids: regular, 1ff; deltahedra, 3; Archimedean, 46; Archimedean duals, 50ff; prisms and antiprisms, 61ff
Corners, duality to faces, 4ff; description of, 41; truncation of, 40, 57; star, 68ff, 81, 117ff; invisible, 121; *see also* Faceting
Cross-sections, regular polygonal, 20ff, 177
Cube: as Platonic solid, 1; duality to octahedron, 4, 5; axes of rotational symmetry, 10ff; planes of reflection symmetry, 14, 16; inscribed in sphere, 16; circumscribed to sphere, 20; hexagonal cross-sections of, 23; inscribed in icosahedron or dodecahedron, 32ff; interpenetrating, 34ff; truncation of corners of, 40, 57; snub, 45; dual of snub, 52; dual of truncated, 55; truncation of edges of, 56, 58; faceting, 94; faceting truncated, 99; drilled, 147ff; space filling with, 154; rods supplementing array of, 161; fusion into trefoil knot, 184
Cuboctahedron, 42; dual of, 50; faceting, 96, 125; supplementing drilled cube, 149; tower of, 177

Decorated solids, 137ff
Deltahedra, convex, 3

Descartes, René, 154
Dice: Etruscan, 1; Archimedean duals, 53; dipyramids and trapezohedra, 62
Dipyramids: dual to convex semiregular prisms, 61; dual to star-polygonal prisms, 81
Dodecadodecahedron, 103; dual of, 119ff
Dodecahedron, regular, as Platonic solid, 1; duality of regular to icosahedron, 6; rhombic, 9, 50, 56, 63; axes of rotational symmetry of regular, 13; planes of reflection symmetry of regular, 18; regular inscribed in sphere, 18; regular, circumscribed to sphere, 20; decagonal cross-sections of regular, 24; hexagonal cross-sections of regular, 26; truncated, 44; snub, 45; triangular, 51; dual of snub, 53; dual of truncated, 55; laminated rhombic, 63; Kepler's small stellated, 66, 84; Kepler's great stellated, 67, 85, 87; Poinsot's great, 68, 85, 86; non-regular stellations of regular, 91; faceting regular, 98, 100ff; truncating great, 102ff; stellated rhombic, 134ff; decorated great stellated, 137; decorated rhombic, 140; inside of great, 142; filling space with rhombic, 155; filling space with stellated rhombic, 156, 159; rods supplementing rhombic, 159; rings of rhombic, 179ff; partly interpenetrating rhombic, 181
Duality: of Platonic solids, 4ff; of edges, 8, 51, 72ff, 81; as intellectual instrument, 20; as directive for inscriptions, 29; application to Archimedeans, 50ff; application to convex prisms and antiprisms, 61, 62; of Kepler-Poinsot solids, 71ff; application to star-polygonal prisms and antiprisms, 81; application to non-convex semiregular solids, 116ff

Edges: of deltahedra, 3; of dual figures, 8, 51, 72ff, 81; five groups of, in dodecahedron, 32; truncation of, 56, 58
Etruscans, 1
Euclid, 1

Faces: duality to corners, 4ff; regular star-polygonal, 66ff
Faceting: small stellated dodecahedron, 94; cube, 94; octahedron, 95; cuboctahedron, 96, 125; icosidodecahedron, 96; small rhombicuboctahedron, 97, 110; small rhombicosidodecahedron, 97, 111; truncated cube, 99, 108, 123; regular dodecahedron, 98, 100ff; truncated dodecahedron, 109

Geometric series, 177ff
Growth forms, 178

Heptahedron, semiregular, 95
Heron of Alexandria, 46
Hexakis icosahedron, 55
Hexakis octahedron, 55

Hexecontahedron, pentagonal, 53; trapezoidal, 55
Holes: drilled in solids, 147ff; in arrays of solids, 158ff, 186

Icosahedron: as Platonic solid, 1; duality to regular dodecahedron, 6; axes of rotational symmetry, 13; planes of reflection symmetry, 18; inscribed in sphere, 18; circumscribed to sphere, 20; decagonal cross-sections of, 24; irregular dodecagonal cross-sections of, 27; truncated, 43; dual of truncated, 55; Poinsot's great, 69, 93ff; stellation of, 92ff, 117; truncating great, 104
Icosidodecahedron, 42; dual of, 50; faceting, 96, 103; great, 112ff; faceting great, 114ff; dual of great, 117
Icositetrahedron: pentagonal, 52; trapezoidal, 55
Insides, 121, 142ff; of insides, 151
Interpenetrating lattices, 166, 173
Interpenetrating solids: Platonic duals, 9; two cubes, 34, 78; three cubes, 35; four cubes, 35; five cubes, 36; two dodecahedra, 37; five tetrahedra, 38ff; capped by star-polygons, 78ff; two truncated tetrahedra, 123; three star-octagonal prisms, 123; three square prisms, 147, 149; *see also* Partial interpenetration
Isometry, 126ff, 132, 154ff, 165, 167

Kepler, Johann, 61, 66ff, 82
Knot: with three crossings, 184ff; with four crossings, 187

Laminated solids, 63, 141
Lattices: simple cubic, 162; face-centered cubic, 163; body-centered cubic, 164; interpenetrating, 166; unit cells, 168; primitive cells, 170; distinguished from structures, 173; fathering trefoil knots, 184ff
Left-right distinction, *see* Right-left

Making shapes: of cardboard, 188ff; of wire, 192
Mirror image, *see* Planes of reflection symmetry
Monsters, 28, 49

Nine regular solids, 70
Nolids: regular isometric, 124, 138ff; defining linked rings, 182ff
Non-performable operations, 20, 30, 47; *see also* Right-left distinction

Octahedron: as Platonic solid, 1; duality to cube, 4, 5; axes of rotational symmetry, 10ff; planes of reflection symmetry, 15, 17; inscribed in sphere, 15; hexagonal cross-sections of, 20; truncation of corners of, 40; dual of truncated, 55; stellating, 82; faceting, 95; tetrahedrally colored, 137; supplementing drilled tetrahedron, 150; supplementing tetrahedral array, 157; four fused rings of, 179; fusion into trefoil knot, 185
Operations, consistent symmetry, 127, 132

Pappus of Alexandria, 46
Partial interpenetration: tetrahedra, 175; cuboctahedra, 177; truncated octahedra, 177; great stellated dodecahedra, 178; rhombic dodecahedra, 181
Pentagram, 65ff
Pentakis dodecahedron, 55
Performable operations, 20; *see also* Right-left distinction
Planes of reflection symmetry, 14ff, 20
Plato, 1, 46
Poinsot, Louis, 68
Polygons, *see* Regular polygons
Primitive cells, 170
Prisms: convex semiregular, 61; dipyramids dual to, 61; star-polygonal semiregular, 76ff; duals to star-polygonal, 81; interpenetrating star-octagonal, 123
Pythagoreans, 65

Reflection, *see* Planes of reflection
Regular polygons: convex, 1; as cross-sections, 20ff; as faces on semiregular solids, 41; non-convex, 65, 75, 79
Regular solids: Platonic, 1ff; axes of rotational symmetry of, 10ff; planes of reflection symmetry of, 14ff; inscribed and circumscribed to spheres, 20; cross-sections of, 20ff; defined, 41; non-convex, 66ff; Kepler-Poinsot, 70
Rhombicosidodecahedron: small, 43; great, 44; duals of small and great, 55; faceting small, 97, 111
Rhombic solids, 9; *see also* Cube, Dodecahedron, Triacontahedron
Rhombicuboctahedron: small, 43, 105; great, 44; duals of small and great, 55; faceting small, 97, 110; drilled great, 152ff
Right-left distinction, 20, 47ff, 127ff; tetrahedron in dodecahedron, 30ff, 39; snub solids, 45; duals to snub solids, 53; edge-truncated cube, 59; stellated regular dodecahedron, 91; regular triangular nolid, 125, 138ff, 182; stellated rhombic dodecahedron, 135, 159; decorated solids, 140ff; families of paths through space, 159ff; interlinked rings, 182; trefoil knots, 186
Rings: of fused rhombic dodecahedra, 179; of fused octahedra, 179; of interpenetrating rhombic dodecahedra, 181; linked regular polygonal, 182ff
Rods: isometric assemblies of, 159ff; triangular, supplementing rhombic dodecahedra, 159; square, supplementing cubes, 161

Semiregular solids: defined, 41; inscribable in spheres, 41; convex, 42ff; Archimedean, 46; prisms and antiprisms, 61, 76, 80; non-convex, 95ff; duals to non-convex, 116ff
Series, geometric, 177ff
Snub solids, 45; duals of, 52ff

Space, non-intersecting paths through, 160ff
Space filling: cubes, 154, 162; rhombic dodecahedra, 155, 163; truncated octahedra, 155, 164; stellated rhombic dodecahedra, 156, 159, 165; tetrahedra and octahedra, 157; rhombic triacontahedra and supplements, 158, 166; rhombic dodecahedra and rods, 159; cubes and rods, 161; in diamond structure, 174
Sphere: inscribed and circumscribed to regular solid, 20; circumscribed to semiregular solid, 41, 116; inscribed to Archimedean dual, 51; approximation to, 53, 194; inscribed to non-convex semiregular dual, 117; packing of repeated, 163
Star of David, 75, 78, 83
Star polygons, 65, 75, 79
Stella octangula: as interpenetrating tetrahedra, 9; as stellation of octahedron, 82; as faceted cube, 94; truncated, 123; repetitive array of, 157
Stellating: octahedron, 82; truncated octahedron, 83; regular dodecahedron, 84ff; icosahedron, 92ff, 117; rhombic triacontahedron, 117ff; rhombic dodecahedron, 134ff
Structures, diamond, 173; solids associated with, 174
Supplements: to drilled solids, 147ff; to repetitive arrays, 158ff, 166
Symmetry, *see* Axes, Planes, Isometry, Right-left

Tetrahedron: as Platonic solid, 1; self-duality, 7; compounds of, 9, 38ff, 82; axes of rotational symmetry, 13, 19; planes of reflection symmetry, 19; inscribed in cube or octahedron, 19, 29; square cross-sections of, 20; inscribed in regular dodecahedron, 30, 48; inscribed in icosahedron, 31; truncation of, 42; drilled, 150; supplementing octahedral array, 157; partly interpenetrating, 175
Tetrakis hexahedron, 55
Theaetetus, 1
Towers, 10ff, 177ff
Trapezohedra: Archimedean, 54; dual to convex antiprisms, 62; dual to star-polygonal antiprisms, 81
Trefoil knots: of fused cubes, 184; of fused octahedra, 185; assembly of four, 186
Triacontahedron: rhombic, 9, 50, 57; stellation of, 117ff; repetitive array of, 158
Triakis icosahedron, 55
Triakis octahedron, 55
Triakis tetrahedron, 51
Truncated: tetrahedron, 42, 150; octahedron, 43, 83, 155, 164, 177; icosahedron, 43, 172; cube, 44, 99, 108; dodecahedron, 44, 109; great dodecahedron, 102; great icosahedron, 104
Truncation: of corners, 40, 57; of edges, 56

Unit cells, 168